ACADIA 2022

Akbarzadeh
Aviv
Jamelle
Stuart-Smith

Hybrids & Haecceities

EDITORS
Dr. Masoud Akbarzadeh, Dr. Dorit Aviv, Hina Jamelle, Robert Stuart-Smith

COPYEDITOR
Gabi Sarhos

GRAPHIC IDENTITY
Madison Green, Paul Germaine McCoy, Peik Shelton

LAYOUT DESIGN
Anna Ji-Eun Lim

PRINTER
IngramSpark

Conference hosted by the University of Pennsylvania, Weitzman School of Design in Philadelphia, Pennsylvania on October 27th - 29th, 2022.

ISBN 979-8-9860805-7-4

Hybrids & Haecceities

The Projects Catalog

2022

The Projects Catalog for the 42nd Annual Conference of
the Association for Computer Aided Design in Architecture

Weitzman School of Design at the University of Pennsylvania

Hybrids & Haecceities

Table of Contents

INTRODUCTION

FEATURED EXHIBITION

RESEARCH PROJECTS

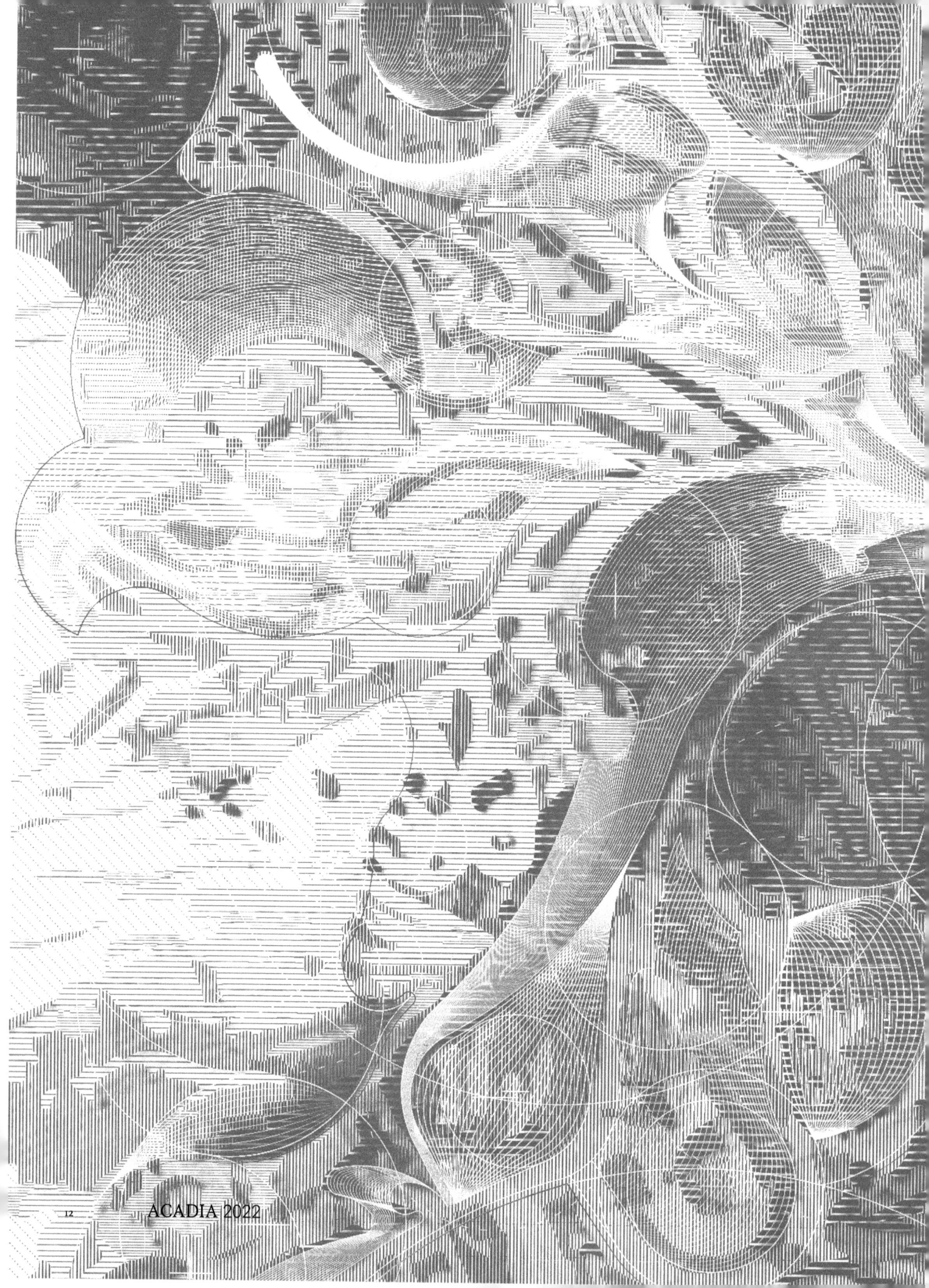

CREDITS

Hybrids & Haecceties

Towards Diverse Forms of Participation in Design

Dr. Masoud Akbarzadeh, Dr. Dorit Aviv, Hina Jamelle, Robert Stuart-Smith

Hybrids & Haecceities provides a unique opportunity to reflect on the challenges and opportunities in undertaking computation-led research, practice, or theory in design today. During the last few years alone, the US and global communities have experienced considerable social, economic, environmental, and political upheaval, in an increasingly polarized society that struggles to achieve equity. Many of these issues are further exacerbated by the ever-more apparent effects of climate change.

In this context, the design and construction industries continue to have profound global effects with significant political, economic, and environmental consequences. Today's social and ecological crises are vast and complex, however, and cannot be addressed through design alone. To confront these issues, designers must forge new collaborations, and extend their means of analysis, deliberation and action, furthering design agency. Perhaps one of the greatest challenges we must face is the fact that design opportunities are not easily integrated within computational paradigms and, when generalized, can have profound impacts. Joy Buolamwini's research into algorithmic bias highlights how even simple oversights in the creation of narrow data sets can easily lead to biases that can have broad and unintended impacts such as unfair racial profiling in policing and law enforcement (Buolamwini and Gebru 2018). Cathy O'Neil describes such technologies as "weapons of math destruction" (WMDs), where overly simplified heuristics within computational models prevalent in almost every aspect of society have led to exclusionary or inequitable treatment. O'Neil suggests that most of today's computational models fail to sufficiently represent, describe or include the diversity of today's world and communities (O'Neil 2016).

Since the middle of the twentieth century, pioneers in the architecture and design professions have sought to leverage computing to integrate diverse performances and effects in the design of the built environment. Nicholas Negroponte's Architecture Machine Group (Negroponte 1972) and others developed computational and robotics research in the early 1950s, while Cedric Price together with John and Julia Frazer, explored the integration of computing in speculative building proposals to support user flexibility and play (Frazer 1995). Several Viennese architects in the 60s and 70s utilized digital media and sensor-feedback to support a wearable, event-based architecture. Together these explorations could be aligned to Donna Harraway's *Cyborg Manifesto*, if they were directed more towards empowerment and eclecticism (Haraway and Wolfe 2016). The digital design movement of the 1990s and early 2000s sought to extend the designer's abilities to address complex and diverse architectural considerations by leveraging computing for both performance modelling, and its inherent ability to produce geometric variability. Since the turn of the century this endeavour has turned increasingly towards material concerns. Beyond the automation of tasks, many designers explored the encoding of degrees of difference in families of buildings, building elements, and in material composition through approaches to simulation, design, or fabrication. Computation offered a means to address variable conditions and produce variable outcomes or outcomes that embody internal variation.

Variability, however, does not necessarily provide considered responses to different user groups, environments, or site scenarios. Computation and design methods can easily generalize a design problem, as can be seen when a suite of projects follow shortly after a new software plugin or algorithm is made publicly available. For variability to be meaningful, it might engage in the specifics of each project and its stakeholders, and offer bespoke, tailored, site-specific, personalized, or user-customized solutions. Until recently, such one-off design outcomes would have been cost-prohibitive to produce, and the computational models arguably not sufficiently robust for deep levels of customization to avert the potential pitfalls of generalization.

However, with the emerging Fourth Industrial Revolution (Industry 4.0), a fundamental shift away from abstract generalized models of mass production is presently taking place, toward greater degrees of customization at unprecedented scales. Its key enabling technologies include autonomous robotics, autonomous manufacturing, and 3D printing, among others. These are supported by cyber-physical systems, internet of things (IoT) technologies, decentralized infrastructure (Schwab 2017), and are also increasingly being augmented by deep learning models. The vast interconnectedness of these technologies will be disruptive and potentially risk an exponential rise in production, consumption and waste that poses an existential threat to not only our climate and its delicately balanced ecosystems, but also to social equity. There is also an urgent need to decarbonize buildings, to reduce cost and increase production of housing to meet demand, and to provide equitable infrastructure to communities at risk. Concurrently, however, Industry 4.0 technologies also support a shift to more diverse and considered forms of embodiment and participation in the built environment.

How might alternative models of design address our diverse world? Can we reflect our vast genders, races, ethnicities, species, landscapes, bio-synthetic or genetically engineered conditions, and augmented worlds within our designs? How could one cater for, or engage meaningfully with such diversity? This year's theme, *Hybrids & Haecceities,* seeks to contribute ideas to this question. The theme's use of these two terms comes with a call to action to critically and creatively reassess the means in which we conceive, undertake, and implement design. Both "hybrids" and "haecceities" carry several associations that operate together to provide fertile ground for framing future design work and discourse.

Hybrids are entities with characteristics enhanced by the process of combining two or more elements with different properties. Bred hybrids such as mules or ligers (horse-donkey and lion-tiger hybrids) embody different physical and behavioral characteristics from their parents. A liger might not only appear to be something between a lion and a tiger, but also inherit a tiger's swimming capabilities and a lion's social skills. More surprisingly though, the mix of the two species' genes can produce characteristics not apparent in either parent species—ligers are typically larger than lions or tigers (Bonnicksen 2009). Hybrids can exist in varying ratios and be more than their constituent parts. Hybrids, therefore, dispel binary thinking and offer opportunities for the combining of materials (such as novel composites), technologies, or objects for a valuable outcome. An interest in hybrids is a rejection in dualistic binary conditions, and asserts new forms of unity that might address ethical, social, theoretical, or environmental concerns, or be a means to approach engineering, design, or aesthetics. Hybridization can lead to untold possibilities.

The first known use of word *Haecceity* is attributed to Philosopher-theologian-priest John Duns Scotus in thirteenth century CE to describe a non-qualitative property that defines something as unique and indivisible. For Scotus, haecceity was essentially the "thisness" (derived from the Latin "haec" meaning "this") as opposed to a "whatness" or essence of an object or substance (Cross 2022). In *The Rise of Realism*, philosophers Manuel Delanda and Graham Harman agree that haecceities describe uniquely identifying features, or "thisness" of a specific cat, as opposed to its breed or species that equally describe several different cats. They differ on whether haecceities also involve history, generative processes, artefact, or agency (DeLanda and Harman 2018). *Haecceities* as a term, challenges a generalization of things by focusing on what is unique.

In concert, *Hybrids & Haecceities* offers a provocation towards more inclusive and specific forms of computational design. *Hybrids & Haecceities* rejects binary thinking at all levels, is critically optimistic towards technology, and seeks more open and diverse engagement with the world at large, where each engagement is specific. *Hybrids & Haecceities* can and should also be interpreted as an agenda that seeks more considered design outcomes, that lead us towards new impactful solutions leveraged by new technologies.

The *ACADIA 2022 Hybrids & Haecceities: The Projects Catalog* is the first of two volumes that document the ACADIA 2022 conference. It showcases peer-reviewed project submissions. Presented projects were double-blind peer-reviewed and competitively selected for inclusion in the conference. Project submissions included in this publication were also exhibited during the three days of the conference. Additionally, a selection of work from faculty at the host institution — University of Pennsylvania Weitzman School of Design is included. Also exhibited during the conference, the faculty work responds to the conference brief in diverse ways, as outlined in an essay written by the ACADIA 2022 Exhibition Chair, Ferda Kolatan. Beyond the exhibition, the host institution developed a unique approach to the conference's graphics which adorn the proceedings volumes, conference website and merchandise. A section devoted to this venture provides some insight into this bespoke approach to graphic design.

The *ACADIA 2022 Hybrids & Haecceities: The Projects Catalog* provides an extensive overview of this year's projects and other key activities in the conference. Together with the paper proceedings included in the second volume, the two ACADIA 2022 conference books demonstrate an immense and diverse body of design research that we hope provides inspiration for years to come and ushers in more impactful and exciting work in subsequent years. The high-level of rigor and quality in the work is self-apparent, and provides a substantial contribution towards computational design research.

REFERENCES

Bonnicksen, A. L. 2009. *Chimeras, Hybrids, and Interspecies Research: Politics and Policymaking*. Washington, DC: Georgetown University Press.

Buolamwini, J., and T. Gebru. 2018. "Gender Shades: Intersectional Accuracy Disparities in Commercial Gender Classification." In *Proceedings of the 1st Conference on Fairness, Accountability and Transparency*, PMLR 81: 77-91. https://proceedings.mlr.press/v81/buolamwini18a.html.

Cross, R. 2022. "Medieval Theories of Haecceity," *The Stanford Encyclopedia of Philosophy* (Spring 2022 edition), edited by E. N. Zalta, https://plato.stanford.edu/entries/medieval-haecceity/

DeLanda, M., and G. Harman. 2018. *The Rise of Realism*. New York, NY: John Wiley & Sons.

Frazer, J. 1995. "An Evolutionary Architecture." In *An Evolutionary Architecture*. London, UK: Architectural Association.

Haraway, D. J., and C. Wolfe. 2016. Manifestly Haraway. Minneapolis, MN: University of Minnesota Press.

Negroponte, N. 1972. *The Architecture Machine: Toward a More Human Environment*. Cambridge, MA: MIT Press.

O'Neil, C. 2016. *Weapons of Math Destruction: How Big Data Increases Inequality and Threatens Democracy*. New York, NY: Crown.

Schwab, K. 2017. *The Fourth Industrial Revolution*. New York, NY: Crown Publishing Group.

ACADIA 2022

Featured Exhibition

6   Brooklyn Botanic Garden Visitor Center, Brooklyn
     WEISS/MANFREDI

4   Brooklyn Botanic Garden Visitor Center, Brooklyn
     WEISS/MANFREDI

2   Global Innovation Center, Philadelphia
     Erdy McHenry Architecture, LLC

1   Prototype for Ger, UNICEF, Mongolia

William Braham with Evan Oskierko-Jeznacki and Max Hakkarainen

3   Global Innovation Center, Philadelphia
     Erdy McHenry Architecture, LLC

7   Longwood Gardens, Brandywine Valley
     WEISS/MANFREDI

5   Brooklyn Botanic Garden Visitor Center, Brooklyn
     WEISS/MANFREDI

8   Longwood Gardens, Brandywine Valley
     WEISS/MANFREDI

A Mosaic of Unlikely Affinities

University of Pennsylvania, Weitzman School of Design

Ferda Kolatan

The ACADIA 2022 conference at the University of Pennsylvania Weitzman School of Design includes an exhibition featuring select work from Weitzman's architecture faculty. *Hybrids & Haecceities: A Mosaic of Unlikely Affinities* takes on the larger questions raised by the conference theme and reflects on them through a kaleidoscopic assemblage of images depicting prototypes, artifacts, and buildings produced by individual faculty members or with their respective teams and practices. The ambition for this exhibition is twofold. First, to introduce the host school's diverse design and technology faculty to the wider conference audience and to showcase Weitzman's strong and longstanding commitment to computer aided design as an indispensable driver for architectural research, innovation, and advancement. And second, to locate unexpected kinships and affinities among the exhibited work despite their many differences in style, scale, program, and purpose.

Since the late 1990s, Penn's architecture department has been widely renowned as a leading venue for the exploration and development of digital technologies, tools, and techniques. From the earliest pioneering efforts in surface and mesh modeling, parametric design, and digital fabrication to current research into robotics, AI, and complex material, structural, and environmental systems, the Weitzman School has consistently been pushing the boundaries of our discipline and providing students, researchers, and visiting scholars with an open platform to engage in productive debate and rigorous inquiry.

The faculty show provides a glimpse into these debates and inquiries through the multifaceted work of the instructors, which, in most cases, has developed in parallel to their own teaching and academic research at Penn. After all, the countless innovations brought along by the "digital turn" over the past quarter century, the truly transformative effects they had, and continue to have, on how contemporary architecture is conceived, designed, fabricated, build, and represented, often took shape first as a set of experimental research questions buried deep in an instructor's course syllabus.

The implementation of digital technologies toward architectural practice evolves through a series of intermediary stages starting out in the realm of abstraction, conjecture, hypotheticals, and hunches before slowly moving toward greater robustness as the various constraints of "real" life are confronted one by one. But it is important to recall that all stages in this process are in themselves intricate collaborative acts between students, teachers, peers, critics, and other professionals from related fields as well as nonhuman factors like organic matter, natural forces, machines, protocols, and policies. The totality of these efforts, their compounded value for architecture, is not reducible to individual tasks, problems, or projects. In the end, what we refer to as "novelty," "agency," or "progress" are not discernible properties adorning the final product like a crown but instead qualities emerging in more subtle and labored ways from the collective reservoir of a myriad hard-fought endeavors that painstakingly lay the groundwork for true innovation.

10 497 GW Condominium, New York
Archi-Tectonics

12 Millenium Hall, Philadelphia
Erdy McHenry Architecture, LLC

9 Longwood Gardens, Brandywine Valley
WEISS/MANFREDI

11 497 GW Condominium, New York
Archi-Tectonics

13 Hybrid Main Stadium, Hangzhou
Archi-Tectonics

14 Hybrid Main Stadium, Hangzhou
Archi-Tectonics

15 Hybrid Main Stadium, Hangzhou
Archi-Tectonics

16 Lijia Smart Park Ford Mobility Center, Chongching
Ali Rahim and Hina Jamelle_Contemporary Architecture Practice NY SH

The *Mosaic of Unlikely Affinities* hopes to capture this condition conceptually as well as visually. Along the main wall of the exhibition, a lose directional organization guides the visitors from artifacts, objects, and prototypes to experimental techniques in digital imaging and visual representation and finally to photographs of actual built structures and building components. While this sequencing seemingly suggests a causal relationship between the scale and type of the exhibited work, the mosaic layout in fact challenges an all-too-linear reading based on familiar formats and hierarchically ordered categories such as "research," "prototype," or "building." The alternating play of differently sized panels, their overlapping and crossing into other project domains, and their capacity to stand alone as individual artifacts, out of context and without the need of supporting images, diagrams, or texts, all privilege multiple simultaneous readings of the exhibited pieces.

Perhaps then the most alluring part of the mosaic wall is the element of surprise. By moving away from a more traditional project-based curation, the individual images are now free to form new alliances and engage in cross-categorical dialog. Only that these alliances and dialogs are, of course, not really *new* but rather the prerequisite constituents of the aforementioned totality of research efforts from which the "compounded value" for architecture derives. In that sense the unexpected juxtapositions and unlikely affinities between the different kinds of architectural and material artifacts at display can be interpreted simply as the genuine representations of the day-to-day operations at the intersection of academia and practice.

The intrinsic contradiction captured in the conference title between two types of objects, those that are composite and mixed in nature (hybrids) and those that have a singular, irreducible identity (as described by haecceities) sets the stage for the exhibition as well. In our all-pervasive digital age, hybrids have long assumed new kinds of authenticities, rendering meaningless the once derogatory term "copy." The digital realm has no origin, root, or source, and if it once did, it got irretrievably lost a long time ago. Unlikely affinities rule this domain, creating artifacts that are promiscuous with an outright oppositional stance toward singular origins, interpretations, and expressions. This might be most evident in the current proliferation of text-to-image AI but

is deeply imbedded in the very DNA of the digital project at large. As we further accelerate toward a world where the allusive combinatorial hybrid will thrive and take on new guises, the question of what defines value and meaning in architecture will dramatically shift. What is the haecceity of a hybrid? Does it have one? Does it *want* one? And if the answer to these questions is "no," then what are the mechanisms of assessing the value of digital architecture beyond its capacity to perform well and please our senses aesthetically? Or put in other words, are digital objects ontologically different from other, non-digital objects?

While there is no way of knowing the answers yet, or perhaps ever, architecture - as a cultural enterprise - has always balanced its ontological and epistemological poles, even if elaborate fictions needed to be constructed to imbue matter with "meaning". The ever-diffusing boundaries between the digital and analog realms, and the subsequent concatenation of these realms into new constellations and territories open up unprecedented opportunities not only for the exploration of new geometries, structures, forms, materials, performances, and affects but also for different and better stories. Given the very real, very material urgency of our dire planetary circumstances and the strange, digitally enforced loss of older concepts of originality and rootedness, alternative fictions to those of the past are sorely needed.

In this spirit, the Weitzman faculty exhibition is best understood as an outsized mood board rather than a carefully curated array of individual projects and firm profiles. Each panel in the mosaic may be part of a larger story already unfolding within the diverse architectural practices of the participants. Together, bound by the mosaic itself, the work may cohere in surprisingly compelling ways or articulate unexpected tensions and incongruities. In either case, hybrids may change their appearance, but they are very much an expression of the political, technological, and aesthetic conditions from which they arise. And, quite possibly, they might just be the most adequate conceptual tool to navigate the future.

17 Church of Saint Aloysius, Jackson
Erdy McHenry Architecture, LLC

20 Asian Games Field Hockey Stadium, Hangzhou
Archi-Tectonics

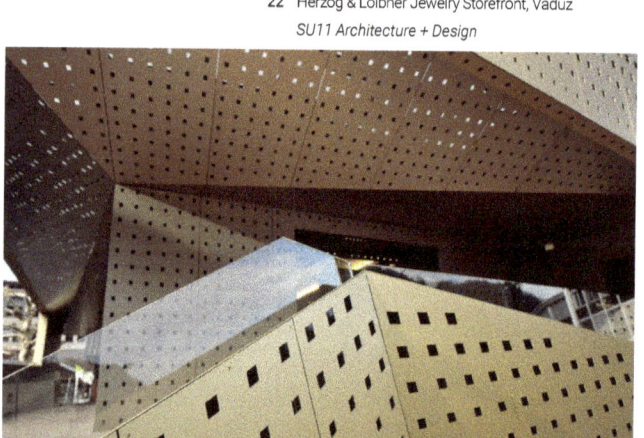

18 Church of Saint Aloysius, Jackson
Erdy McHenry Architecture, LLC

22 Herzog & Loibner Jewelry Storefront, Vaduz
SU11 Architecture + Design

24 Hunter's Point South Waterfront Park, Queens
WEISS/MANFREDI

25 Hunter's Point South Waterfront Park, Queens
WEISS/MANFREDI

23 Herzog & Loibner Jewelry Storefront, Vaduz
SU11 Architecture + Design

19 Church of Saint Aloysius, Jackson
Erdy McHenry Architecture, LLC

21 Asian Games Field Hockey Stadium, Hangzhou
Archi-Tectonics

26 IWI Orthodontics, Tokyo

Ali Rahim and Hina Jamelle_*Contemporary Architecture Practice NY SH*

27 Chia Tai Tianqing Pharmaceutical Headquarters, Nanjing

Ali Rahim and Hina Jamelle_*Contemporary Architecture Practice NY SH*

29 Yale University Tsai Center for Innovative Thinking, New Haven

WEISS/MANFREDI

30 Yale University Tsai Center for Innovative Thinking, New Haven

WEISS/MANFREDI

28 Yale University Tsai Center for Innovative Thinking, New Haven

WEISS/MANFREDI

32 House in a Garden / Garden in a House

Hume Architecture

31 House in a Garden / Garden in a House

Hume Architecture

Hybrids & Haecceities

33 Iceland Green House

Hume Architecture

34 Dekt.io

maeta design LLC

35 Dekt.io

maeta design LLC

36 Mélange

maeta design LLC

37 Lambertucci Inside / Out

A/P Practice

38 Cloud~ing

A/P Practice

40 Tablecloth Compositions

Karel Klein with Shelley Luo and Wenhao Huang

39 Tablecloth Compositions

Karel Klein with Shelley Luo and Wenhao Huang

41 Tablecloth Compositions

Karel Klein with Shelley Luo and Wenhao Huang

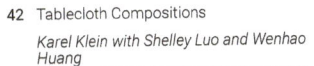

42 Tablecloth Compositions

Karel Klein with Shelley Luo and Wenhao Huang

43 Tablecloth Compositions

Karel Klein with Shelley Luo and Wenhao Huang

44 National Museum of World Writing

Hume Architecture

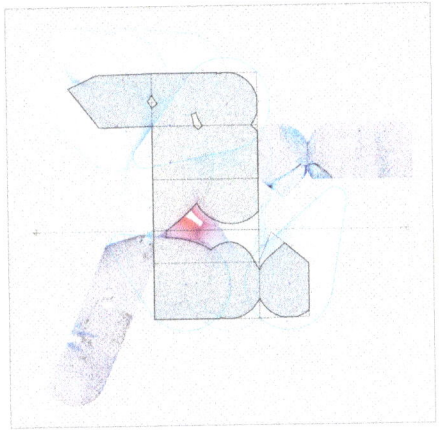

45 House Unfold

Hume Architecture

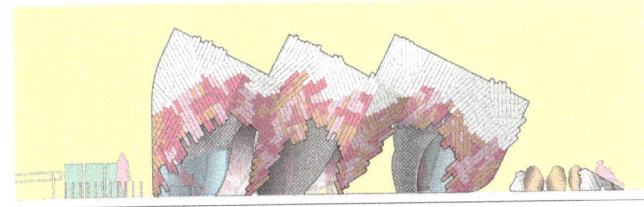

46 Waiting for the Bus

Hume Architecture

47 Ephemeral Images

Hyojin Kwon

48 Ephemeral Images

Hyojin Kwon

49 Untitled
Florencia Pita & Co.

50 Untitled
Florencia Pita & Co.

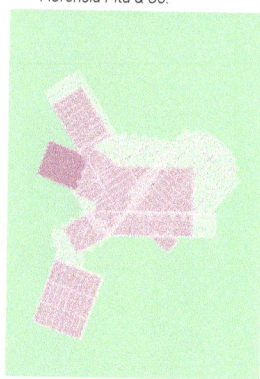

51 Untitled
Florencia Pita & Co.

53 Image as Instruments
Hyojin Kwon

52 Untitled
Florencia Pita & Co.

54 Image as Instruments
Hyojin Kwon

55 Miniature Architecture
Ferda Kolatan with Caleb Ehly

56 Miniature Architecture
Ferda Kolatan with Caleb Ehly

58 Who Are We
Ibañez Kim

60 Who Are We
Ibañez Kim

59 Who Are We
Ibañez Kim

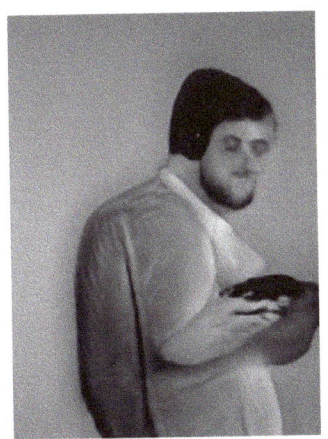

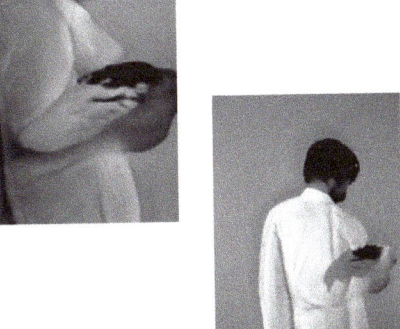

61 Who Are We
Ibañez Kim

62 Who Are We
Ibañez Kim

63 Who Are We
Ibañez Kim

Hybrids & Haecceities   31

64 Who Are We
Ibañez Kim

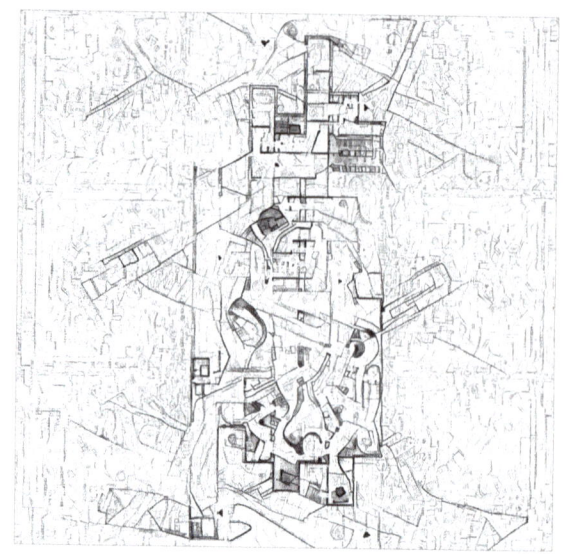

66 Proto-Plans
FORMA Architects PLLC

65 Proto-Plans
FORMA Architects PLLC

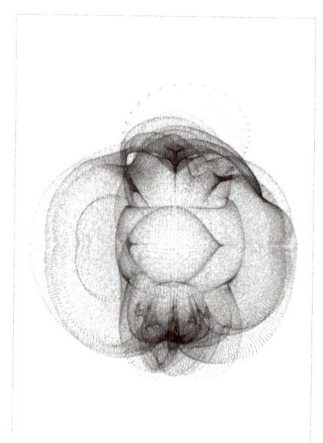

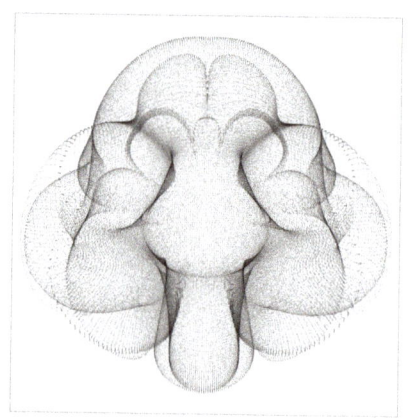

67 A Calatog of Difference
Andrew Lucia

68 A Calatog of Difference
Andrew Lucia

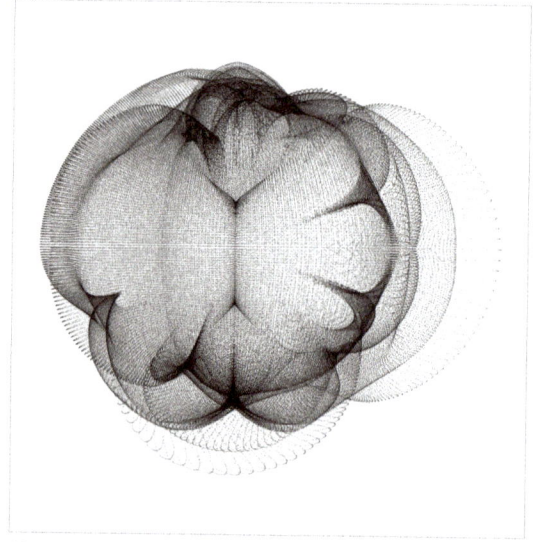

70 A Calatog of Difference
Andrew Lucia

69 A Calatog of Difference
Andrew Lucia

71 A Calatog of Difference
Andrew Lucia

73 Baroque Topologies
Andrew Saunders

72 A Calatog of Difference
Andrew Lucia

74 Baroque Topologies
Andrew Saunders

75 Baroque Topologies
Andrew Saunders

76 Baroque Topologies
Andrew Saunders

77 Prototypes
Robert Stuart-Smith, Autonomous Manufacturing Lab

Hybrids & Haecceities

33

79 Simulation

Autonomous Manufacturing Lab

78 CAAM Speculative Tower

Robert Stuart-Smith, Chris Williams, Paul Shephard

80 Glass Bridge

Masoud Akbarzadeh, Polyhedral Structures Lab

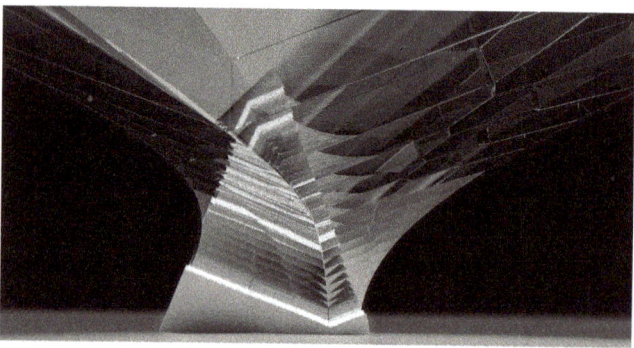

81 Glass Bridge

Masoud Akbarzadeh, Polyhedral Structures Lab

82 Deep Relief

Andrew Saunders

83 Deep Relief

Andrew Saunders

84 Deep Relief

Andrew Saunders

85 Polyhedral Structures

Masoud Akbarzadeh, Polyhedral Structures Lab

87 Polyhedral Structures

Masoud Akbarzadeh, Polyhedral Structures Lab

86 Polyhedral Structures

Masoud Akbarzadeh, Polyhedral Structures Lab

88 Polyhedral Structures

Masoud Akbarzadeh, Polyhedral Structures Lab

89 Hydroculus
Dorit Aviv, Forrest Meggers, Aletheia Ida

90 Hydroculus
Dorit Aviv, Forrest Meggers, Aletheia Ida

91 Hydroculus
Dorit Aviv, Forrest Meggers, Aletheia Ida

92 Hydroculus
Dorit Aviv, Forrest Meggers, Aletheia Ida

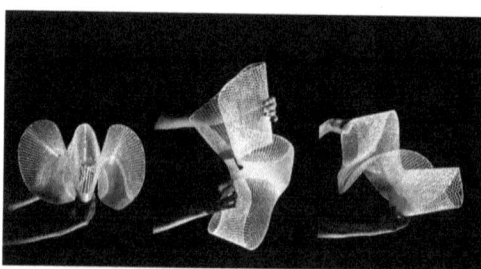

93 Silvanus Series 1
DumoLab

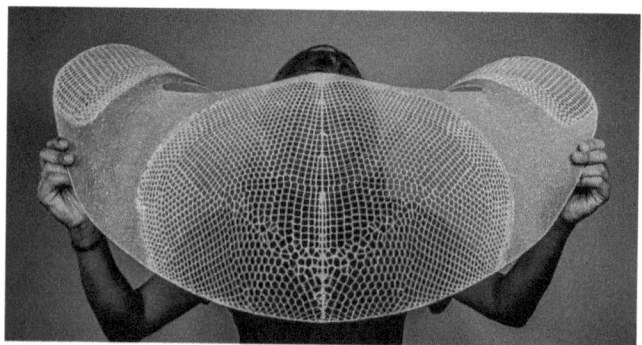

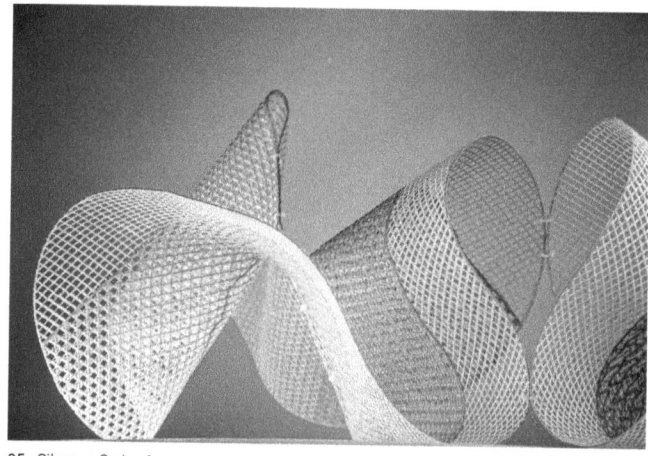

95 Silvanus Series 1
DumoLab

94 Silvanus Series 1
DumoLab

96 Silvanus Series 1
DumoLab

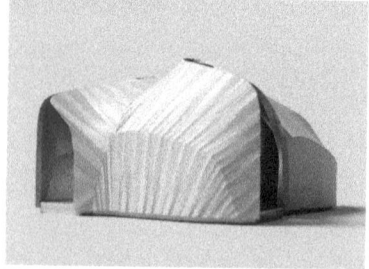

98 Siteless House

Masoud Akbarzadeh, Robert Stuart-Smith, Polyhedral Structures Lab Autonomous Manufacturing Lab, and Polyhedral Structures Lab

97 Siteless House

Masoud Akbarzadeh, Robert Stuart-Smith, Polyhedral Structures Lab Autonomous Manufacturing Lab, and Polyhedral Structures Lab

100 Nanotectonica Metal Relief
Jonas Coersmeier, James Nanasca, Ivan yan Man Hin, Ezio Blasetti

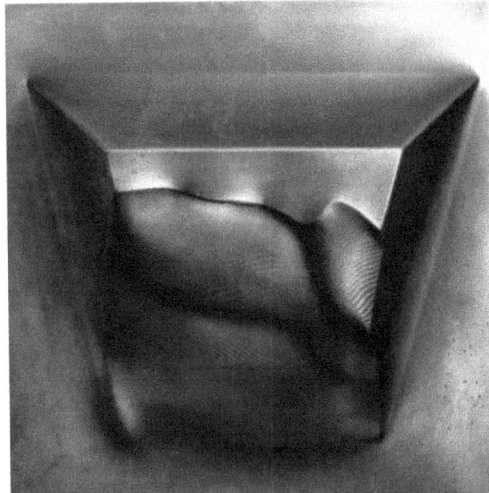

101 Coral Column
SU11 Architecture + Design, NY

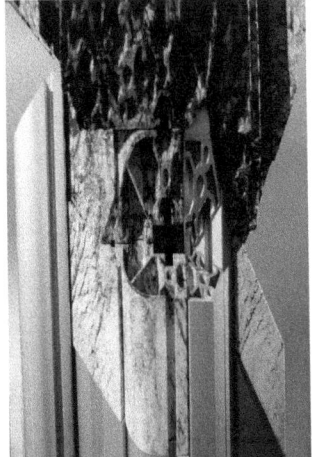

99 Nanotectonica Metal Relief
Jonas Coersmeier, James Nanasca, Ivan yan Man Hin, Ezio Blasetti

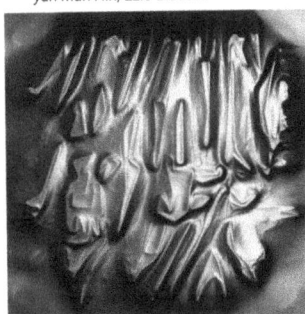

103 Coral Column
SU11 Architecture + Design, NY

104 Coral Column
SU11 Architecture + Design, NY

102 Coral Column
SU11 Architecture + Design, NY

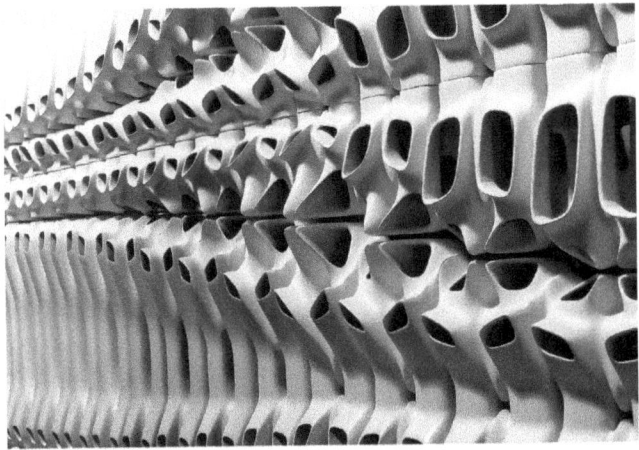

106 Wall of the Future, Museum of Modern Art [MoMA], New York
Ali Rahim and Hina Jamelle_Contemporary Architecture Practice NY SH

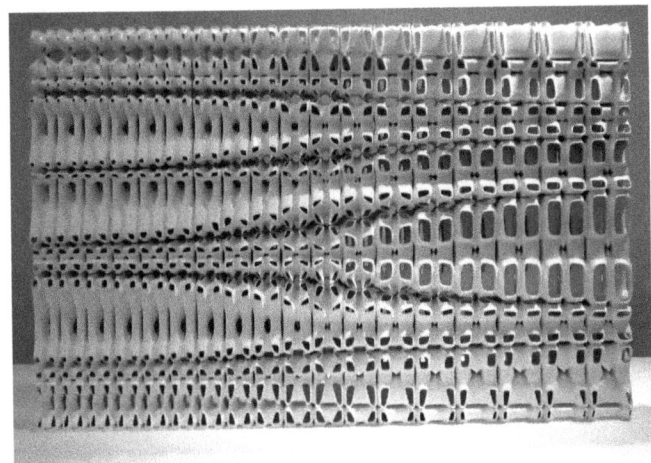

105 Wall of the Future, Museum of Modern Art [MoMA], New York
Ali Rahim and Hina Jamelle_Contemporary Architecture Practice NY SH

Hybrids & Haecceities

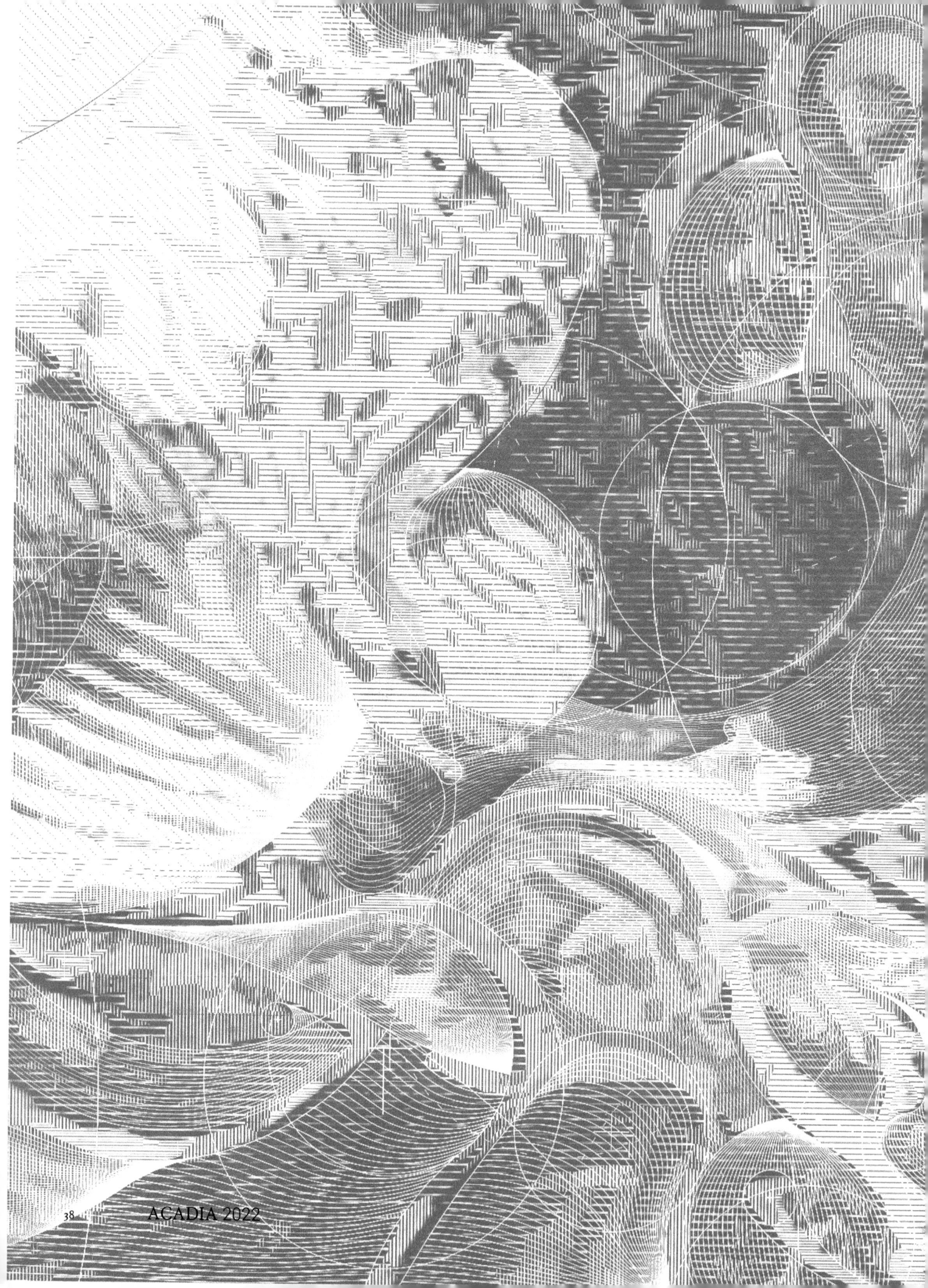

The Research Projects

Degrees of Life

Human-Bacteria Interaction in Architectural Space

Daniela Mitterberger, Tiziano Derme, Barbara Imhof

Degrees of Life is a responsive environment exhibited in February 2022 at Zentrum Fokus Forschung in Vienna. The project explored the interaction between humans and living systems at an architectural scale. The research aims to develop interactive environments within an architectural space that learn, grow, and decay in relation to human presence and behavior (Figures 1, 2). The space reflects on the concept of biomediality and biofacts (Karafyllis 2003), the possible applications of living technologies, and human sensory interfaces in architecture (Hauser 2017, Groutars et al. 2022). *Degrees of Life* is the result of a larger artistic research context called *Co-corporeality* that weaves together architectural design, sensor systems, machine learning, and microbiology.

Environmental Setup

The exhibition was articulated around three distinct self-sustaining closed environments, hosting three types of bacteria: Escherichia coli, Sucrofermenta, and Cyanobacteria strains (Figures 3, 4, 5). The three enclosed environments are named according to the bacteria: ECo, SuCr, and CyA. Although the enclosed environments provided the necessary environmental conditions for the bacteria to survive, they relied on human interaction and mechanical actuation to thrive.

Human Interaction System

Human interaction was registered in real-time by a wearable eye-tracking device that recorded the human visitor's local position and pupil gaze direction (Figure 6). The

PRODUCTION NOTES

Architect:	Co-corporeality, MAEID
Status:	Exhibition / Built
Site Area:	91 sq meters
Location:	Zentrum Fokus Forschung
	Rustenschacherallee 2–4
Date:	2022

1 Visitor wearing the eye-tracking device and interacting with environments

2 Responsive environments reacting to human presence and behavior

3 ECo is an enclosed environment hosting *Escherichia coli* bacteria

4 SuCr is an enclosed environment hosting Sucrofermentas bacteria and microbial biomass production

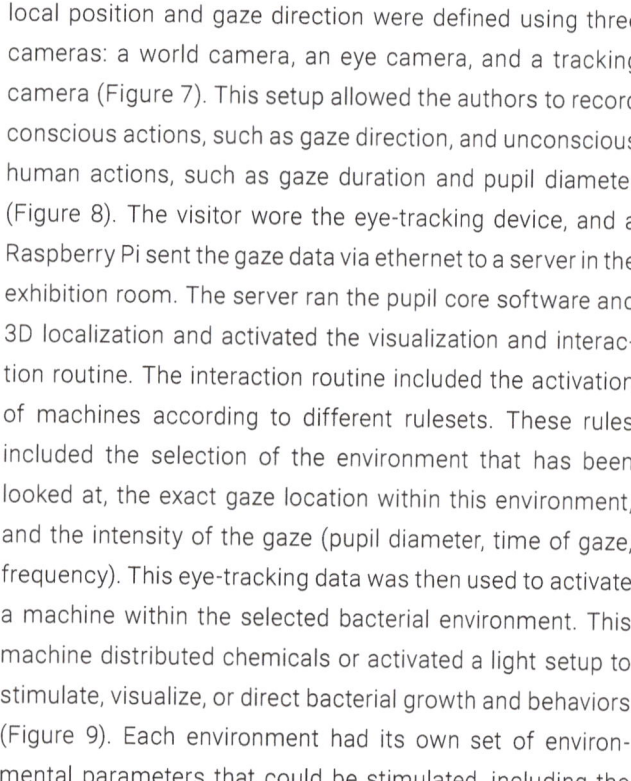

5 CyA environment with Cyanobacteria reacting to different light stimuli

local position and gaze direction were defined using three cameras: a world camera, an eye camera, and a tracking camera (Figure 7). This setup allowed the authors to record conscious actions, such as gaze direction, and unconscious human actions, such as gaze duration and pupil diameter (Figure 8). The visitor wore the eye-tracking device, and a Raspberry Pi sent the gaze data via ethernet to a server in the exhibition room. The server ran the pupil core software and 3D localization and activated the visualization and interaction routine. The interaction routine included the activation of machines according to different rulesets. These rules included the selection of the environment that has been looked at, the exact gaze location within this environment, and the intensity of the gaze (pupil diameter, time of gaze, frequency). This eye-tracking data was then used to activate a machine within the selected bacterial environment. This machine distributed chemicals or activated a light setup to stimulate, visualize, or direct bacterial growth and behaviors (Figure 9). Each environment had its own set of environmental parameters that could be stimulated, including the

chemical setup of the environment, lighting conditions, and the dispersion of nutritional supplements. All three environments were triggered according to the needs of the hosted bacteria.

The ECo-environment

ECo was inhabited by the Escherichia coli (E. coli) bacteria. The metabolic process of E. coli led to a change in culture medium pH level, easily detected using pH-sensitive compounds commonly known as pH indicators. The direction of gaze and the pupil diameter of the visitor activated the machinery distributing specific amounts of sodium hydroxide (NaOH) at a precise point into the liquid glucose medium where E. coli were cultured. The release of NaOH results in real-time reversibility of color change in the medium. After that, the metabolic process of the bacteria slowly changes the pH level again and thus also the color of the medium.

SuCr-environment

The SuCr environment supported the Sucrofementas bacteria strain. The cellulosic bacteria secrete out of its

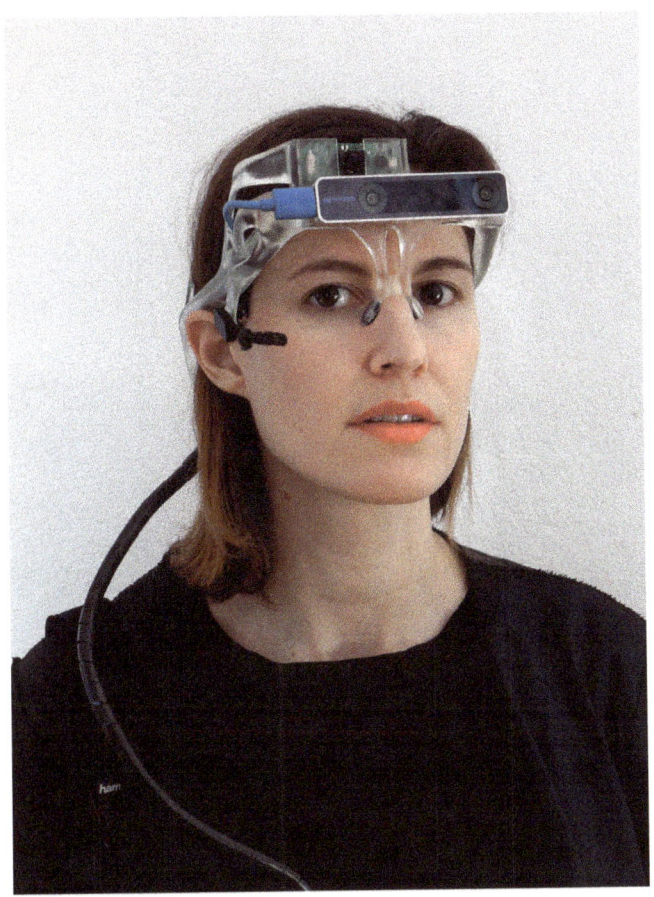

6 Eye-tracking device (prototype 2)

7 Eye-tracking device (prototype 1) detects the gaze direction and position of the visitor

8 Image showing typical interaction of visitors with the ECo environment; visualization projected onto back wall shows the ego-perspective of the visitor

metabolic activity, a thick mat of biomass. The growth of the microbial mat was controlled via a spray nozzle, which moved in two axes and sprayed a nutritional solution (glucose, water, and acetic acid) at a specific location (Figure 10). Human gaze interaction with at specific points in the environment defined the spray location and relatedly the growth rate of the microbial mat.

CyA-environment

The CyA environment hosted the Synechocystis, a genus of cyanobacteria. This bacteria obtains energy via photosynthesis. Human interaction changed the light conditions of the environment by switching the bacteria's growth strategy from photo-autotrophy (light period) to heterotrophy (dark period). The interaction was visualized in real-time by continuously measuring/monitoring dissolved oxygen and pH kinetics. This change in light conditions either activated the photosynthetic activity of the bacteria or reversed it (Figure 11). The environments were placed alongside a visual interface depicting the ego-perspective of the human visitor, the data collected from the visitor's

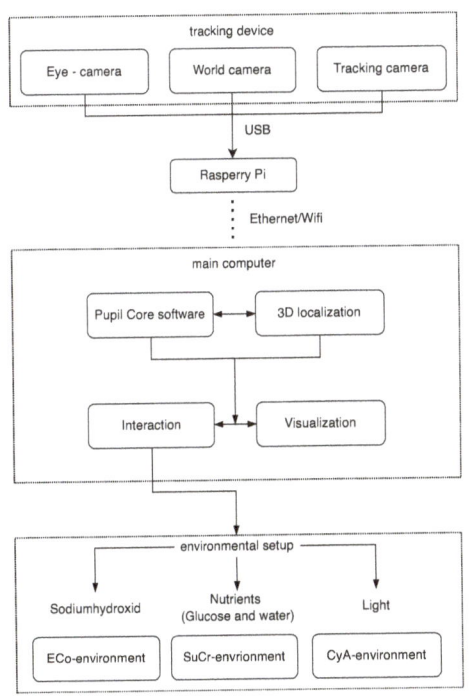

9 Interaction diagram

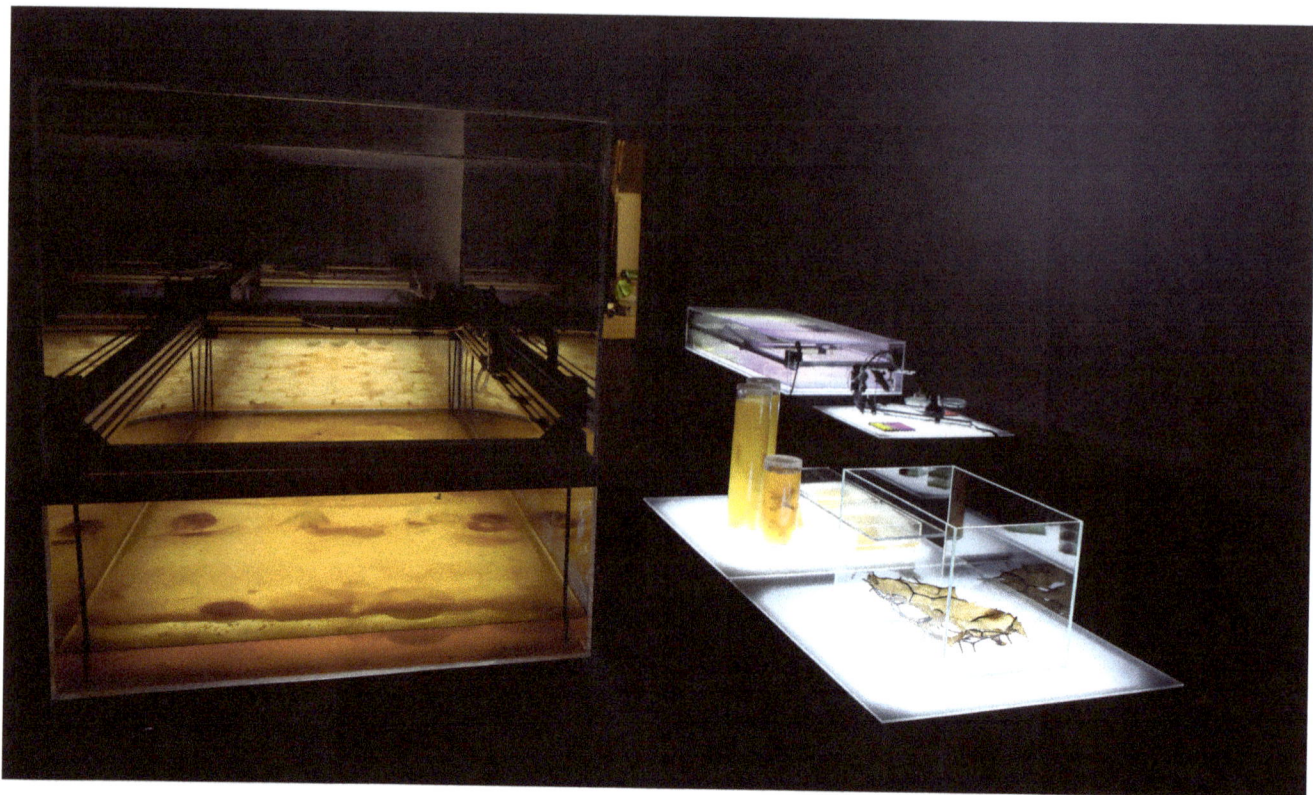

10 SuCr enclosed environment and supplementary microbial prototypes

gaze, and the head position in real-time (Figure 8). A soundscape made the interaction audible and assisted the human visitor in using the system.

Degrees of Life was exhibited for three weeks and evolved depending on the interaction of the human visitors with the environment. The ECo and CyA environments flourished, and no contamination was detected. Conversely, the microbial activity of the SuCr environment, due to its acidity level, accelerated the rusting of the machine's mechanical components. The growth of this environment was primarily automated and only partially attributed to human interaction. The exhibition pursues the idea of interactive architecture as a living system (Maturana and Varela 1980; Beesley 2010), in which physical presence and new modes of observation (Barad 2007) are intertwined with tangible forms of computation (Hauser and Strecker 2020).

ACKNOWLEDGMENTS

The project is part of *Co-corporeality* (AR 00534) and was funded by the Austrian Science Fund (FWF) in the Programme for Arts-based Research (PEEK). We would like to thank all members of the project team and the universities and institutions involved in the scientific and technical development of the research. From the University of Applied Arts, we would like to thank Damjan Minovsky, Nathaniel Loretz, Xavier Madden, Jennifer Cunningham, Patricia Tibu, Kyle Koops, Martin Eichler, and Waltraut Hoheneder. From the Austrian Research Institute for Artificial Intelligence (OFAI), we want to thank Martin Gasser and Robert Trappl. From the University of Vienna, Institute of Materials Chemistry, Polymer & Composites Engineering we want to thank Alexander Bismarck, Neptun Yousefi, Kathrin Weiland, and Anne Zhao. From the University of Vienna, Department of Microbiology and Ecosystem Science, Centre for Microbiology and Environmental System Science we want to thank David Berry and Andreas Heberlein. From the University of Innsbruck, Institute of Microbiology we would like to thank Heribert Insam, Judith Ascher-Jenull, and Carolin Gamirsi. Furthermore, we want to thank our advisory board for their insights and advices along the way, Rachel Armstrong, Alex Arteaga, Philip Beesley, and Petra Gruber. Furthermore we want to thank Zentrum Fokus Forschung for hosting us and Alexander Damianisch and Marianna Mondelos for their support. We want to thank Zita Oberwalder for the photos and Achilleas Xydis for editing the photos.

REFERENCES

Barad, Karen Michelle. 2007. *Meeting the Universe Halfway: Quantum Physics and the Entanglement of Matter and Meaning*. Durham, NC: Duke University Press.

Beesley, Philip, Hayley Isaacs, Pernilla Ohrstedt, and Rob Gorbet, eds. 2010. *Hylozoic Ground: Liminal Responsive Architecture*, 1st ed. Cambridge, Ont.: Riverside Architectural Press.

Groutars, Eduard Georges, Carmen Clarice Risseeuw, Colin Ingham,

Degrees of Life Mitterberger, Derme, Imhof

11 CyA enclosed environment hosting colonies of Synechocystis bacteria (front), ECo environment hosting E.coli bacteria (back)

Raditijo Hamidjaja, Willemijn S. Elkhuizen, Sylvia C. Pont, and Elvin Karana. 2022. "Flavorium: An Exploration of Flavobacteria's Living Aesthetics for Living Color Interfaces." In *CHI '22: Conference on Human Factors in Computing Systems*. New York, NY, USA: Association for Computing Machinery. 1–19. https://doi. org/10.1145/3491102.3517713.

Hauser, Jens. 2017. "Art Between Synthetic Biology and Biohacking: Searching for Media Adequacy in the Epistemological Turn." *Contemporary Arts and Cultures*. Accessed January 15, 2022. https:// contemporaryarts.mit.edu/pub/artbetweensyntheticbiology.

Hauser, Jens, and Lucie Strecker. 2020. "On Microperformativity." *Performance Research* 25(3): 1–7. https://doi.org/10.1080/13528165. 2020.1807739.

Karafyllis, Nicole C., ed. 2003. *Biofakte: Versuch Über Den Menschen Zwischen Artefakt Und Lebewesen*. Paderborn: Mentis.

Maturana, Humberto R., and Francisco J. Varela. 1980. *Autopoiesis and Cognition: The Realization of the Living*. Vol. 42 of the *Boston Studies in the Philosophy and History of Science*. Dordrecht: Springer Netherlands. https://doi.org/10.1007/978-94-009-8947-4.

IMAGE CREDITS

All images by ©Zita Oberwalder.

Daniela Mitterberger is an architect and researcher with a strong interest in new media, the relationship between the Human/Body within digital fabrication, and emerging technologies. She is co-founder and director of MAEID Büro für Architektur und transmediale Kunst, a multidisciplinary architecture practice based in Vienna. Daniela is a PhD researcher and A&T PhD Fellow at ETH Zürich at Gramazio Kohler Research, focusing on human-machine collaboration in digital design and robotic fabrication. Daniela is also a researcher at the University of Applied Arts and co-leader of an FWF PEEK project titled *Co-corporeality*.

Tiziano Derme is an architect interested in the relationship between architectural design, emergent materials, and biotechnologies within digital and robotic fabrication. His research focuses on microbially-mediated fabrication processes applied to the built environment. He is currently a PhD researcher at the Chair for Digital Building Technologies, Institute of Technology in Architecture (ITA) at the Department of Architecture at ETH Zurich. Tiziano is also co-founder and director of MAEID.

Barbara Imhof is a Vienna-based internationally renowned architect, design researcher, and educator. Her projects deal with spaceflight parameters such as living with limited resources, minimal and trans-formable spaces, resource-conserving systems, and all aspects imperative to sustainability. After *Biornametics*, *GrAB-Growing As Building*, *Co-Corporeality* is the third FWF-PEEK funded project she is co-leading. She is the co-founder and co-managing director of LIQUIFER Systems Group (LSG). She has also been teaching at renowned institutes worldwide for over twenty years. Educated in Vienna, London, and Los Angeles, Barbara holds multiple degrees including a PhD.

Disquiet Objects

A Simulated Pensive Domestic Environment

Anirudhan Iyengar

Disquiet Objects is an interactive, Mixed Reality (MR) immersive experience (Figure 1). It follows a hybrid design setup that overlays a physical environment (PE) with a virtual environment (VE) and spatial sound. The project takes place in the setting of a domestic apartment where the PE contains an assemblage of objects, overlaid with a VE that has a completely different visual materiality. It posits an enactive framework—a non-goal-oriented, virtual environment—where the user and the environment constitute the system participants, mediated by a technological artifact. The enactive approach focuses primarily on the dynamic interaction of the organism with its environment, in which perception arises from the connectedness of mind-body-environment through a physical, relational process. The environment is probed through bodily actions, enacting the space beyond the visual modality of representational mapping. The interactive environment aims to question people's objective perception of the domestic space and subjective perception of the self in that space, through the medium of the body.

The experience of walking in the PE, touching the objects, feeling their materiality becomes interactive through the use of an Oculus Quest Head Mounted Display (HMD), which shows completely different qualities of these objects, both

PRODUCTION NOTES

Architect:	Anirudhan Iyengar
Status:	Completed
Location:	aut. Innsbruck, Austria
Date:	2021

1   Disquiet Objects

Hybrids & Haecceities

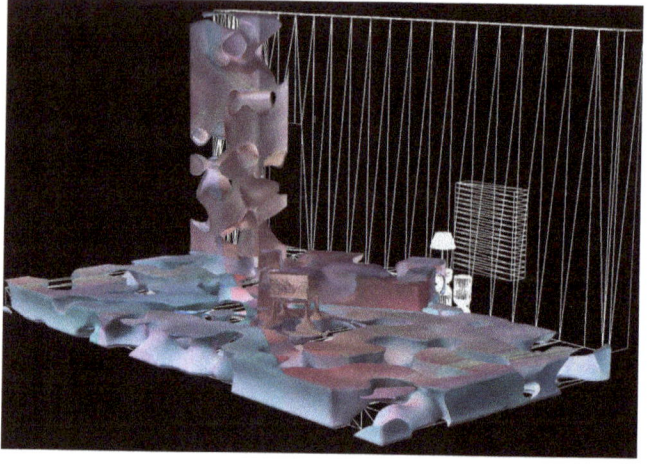

2 Physical Environment setup

visual and haptic. Physically, the person feels and perceives the known materiality of a cushioned chair. However, in the VE, this same chair has a fluidic water-like materiality that merges with the person's avatar. The overlay of VE on PE attempts to create a perceptual disturbance between the visual and the haptic.

This interactive experience split between the PE and VE aims to create a pensive space of embodiment. All the objects in the VE and the person's avatar are rendered with specific materiality, merging into each other, creating a single virtual entity. This entity can be seen either as an extension of the human avatar (Subjective), or as being one with the surrounding environment (Objective). With this action, the project aims to encourage affectiveness in the person, reflecting, questioning, and ideating on their perception of reality (Akoury 2020).

The PE recreates a bedroom space with objects in monotone white, though retaining their inherent material characteristics (Figure 2). The VE in the HMD is calibrated to follow the PE, so that when an object in the PE is touched, it corresponds to the same object in the VE. The human avatar, chair, bed,

3 Virtual Environment setup, depicted with the Raymarching rendering technique. Objects (Signed Distance Field Function) in Raymarching are defined as analytical equations. The absence of polygons or mesh surfaces leads to a greater efficacy in real-time rendering of smooth surfaces and complex, fluidic geometries in the VR HMD.

curtains, and floor in the VE are rendered fluidic, water-like with the same texture, seamlessly blending and merging into one another (Figures 3, 4): a design strategy to elicit confusion in people's perceptual boundaries when differentiating between themselves and surrounding objects.

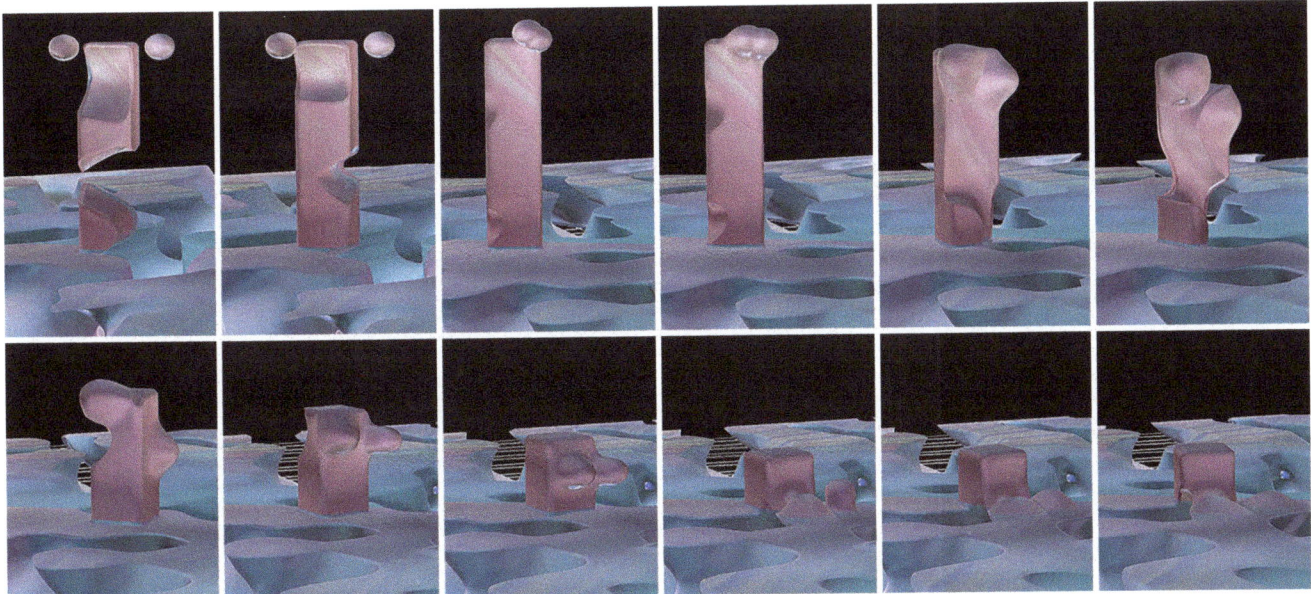

4 Avatar Visualization: the body as a rectangle positioned by the Oculus Quest 2 HMD and hands as spheres facilitated by the controllers; the visitor interacts with their own avatar and the floor in the virtual world, merging into one phenomenal unit.

5 Space interaction diagram of a person

6 Visitor interaction

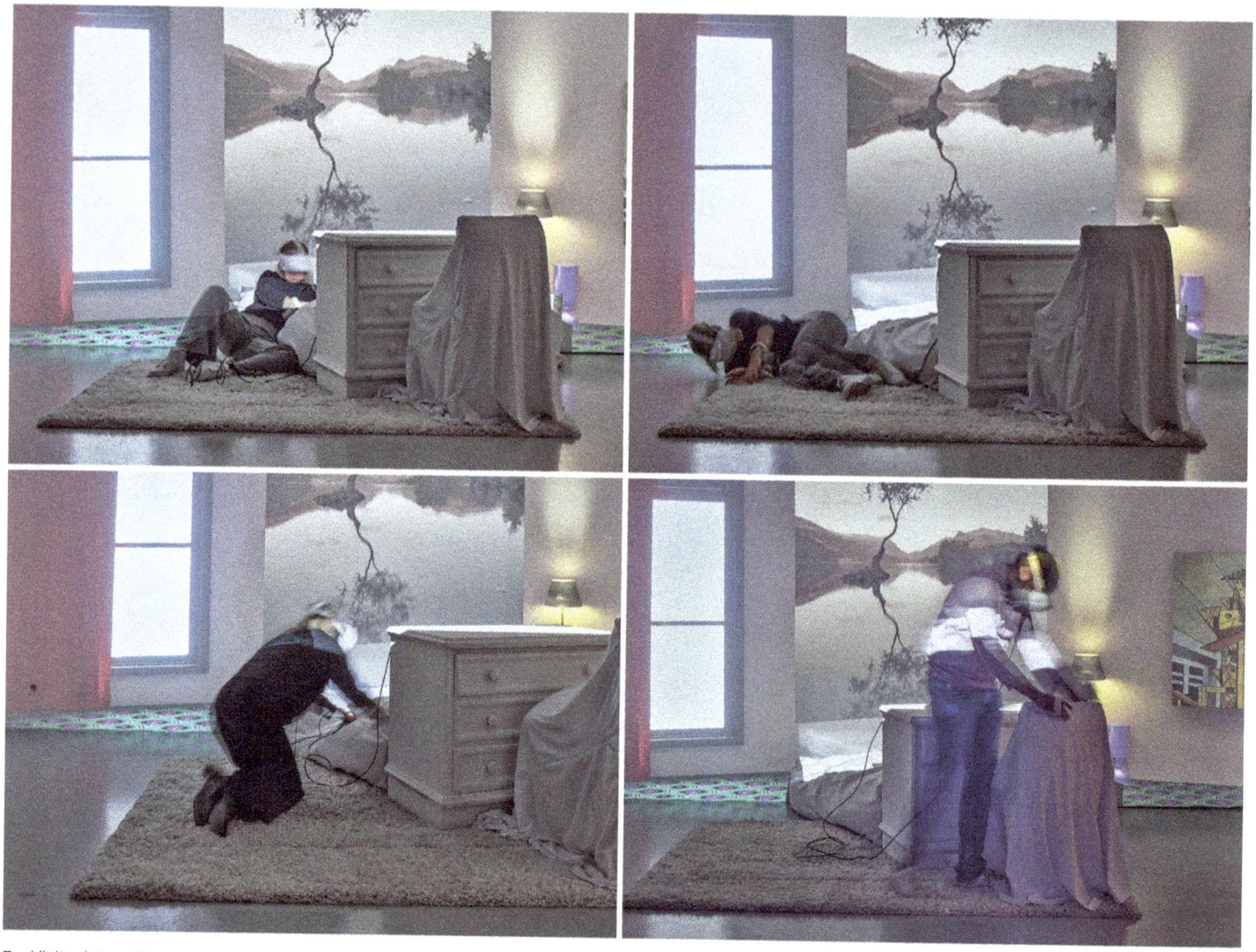

7 Visitor interaction

During the three months the project was on show at an art gallery, visitors had varied reactions to the installation. A few began walking on their hands and knees. Some took over the floor, laying down flat, and interacting with the carpet and bed. Some, seemed possessed by the chair and curtains, obsessively handling them, while some did not want to stop taking part at all (Figures 5, 6, 7). These intense visitor reactions display an affinity with, and comfort derived from the MR environment—a subtle change in the phenomenal ontology underlying visitors' subjective experience (Metzinger 2018).

The illustrated enactive framework challenges the perceptual, cognitive, and operational ontologies of how people engage with their actual domestic setting, concluding that perception and meaning of space arise intrinsically from bodily actions and that technologically mediated, transgressive experiences can bring about a unique perceptual shift in people. The presented body-environment-technology

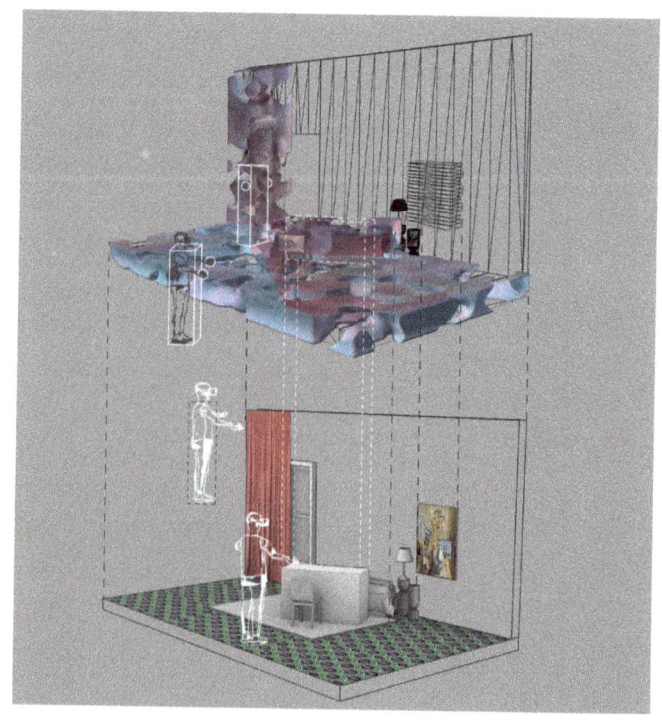

8 Physical Environment overlaid with Virtual Environment and avatar design

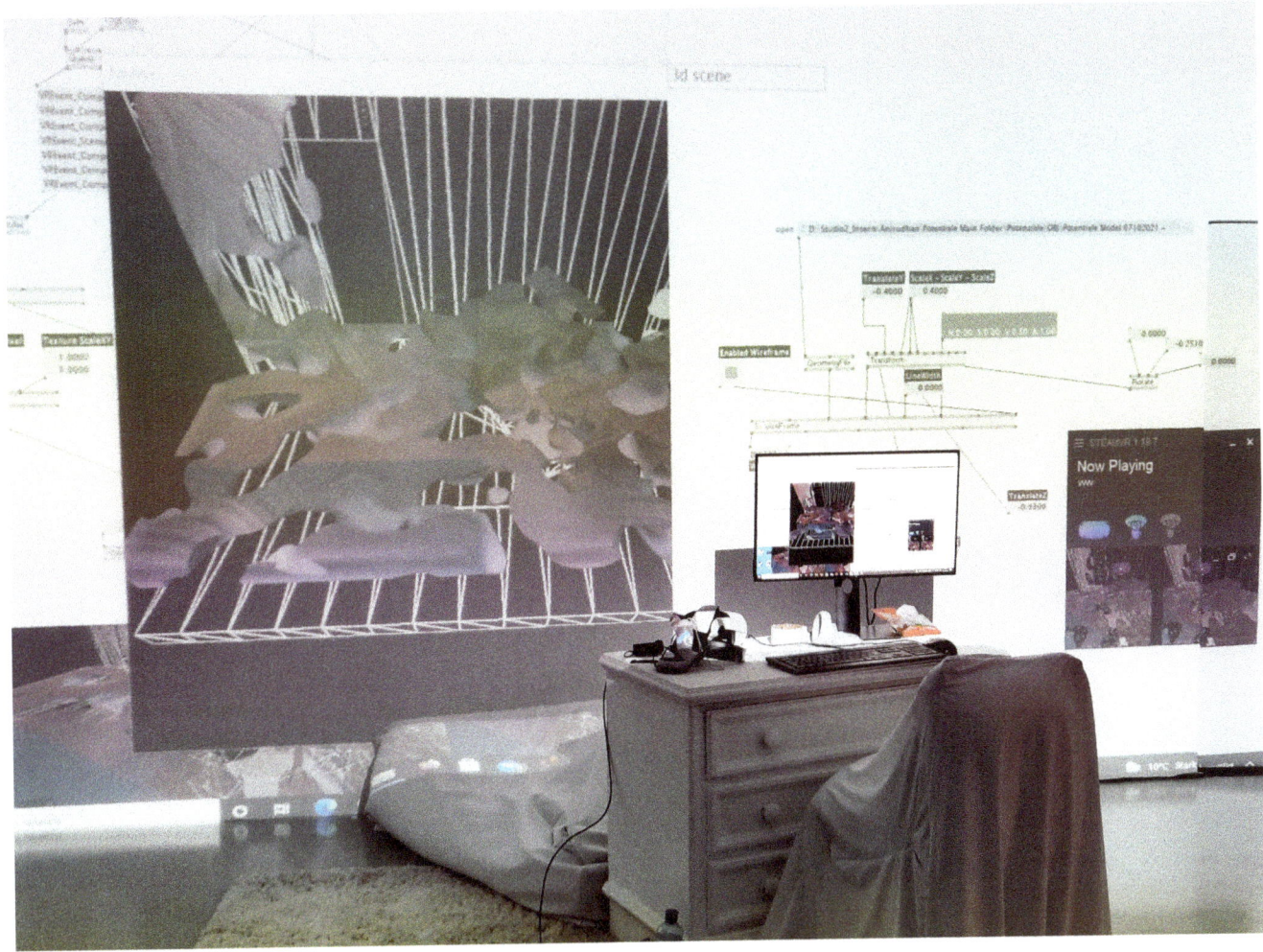

9 Calibration of the virtual environment with the physical environment on site computed in a visual programming application called VVVV

framework can open up new ways to perceive, understand, and design human-environment relationships in the growing technological mediascape.

ACKNOWLEDGMENTS
I would like to thank Clemens Plank and Fiona Zisch for their support and guidance throughout the project and Matthias Holzmann and Ana Maria Stanciu. Thanks also to Studio 2, Institut für Gestaltung, The Faculty of Architecture, UIBK, Gallery aut.architektur und tirol for the opportunity, the funding, and for hosting the exhibition.

REFERENCES
Akoury, Chahid. 2020. "Immersive experiences as the condition of possibility for affective spacing." Continuum 34 (6): 955-963. https://doi.org/10.1080/10304312.2020.1827369.

Metzinger, T.K. 2018. "Why Is Virtual Reality Interesting for Philosophers?" *Frontiers in Robotics and AI* 5. https://doi.org/10.3389/frobt.2018.00101

IMAGE CREDITS
Figures 2, 7: © Günter Richard Wett, 2021
All other drawings and images by the author

Anirudhan Iyengar is an architect, experience designer, and researcher based in Austria. He is currently pursuing a PhD as well as teaching at the University of Innsbruck, Studio2, Institut für Gestaltung. He holds a Master's degree from The Bartlett School of Architecture, UCL and his work has been exhibited internationally at ARS Electronica, The National Museum of China, in London, Moscow, Mumbai, New Delhi & Innsbruck.

LightSense

Architecture as a Creative Partner

Uwe Rieger, Yinan Liu, Tharindu Kaluarachchi, Amit Barde

LightSense is an interactive Extended Reality (XR) installation. It is a design research project that investigates a new generation of responsive architecture by linking a transformable physical construction with its Artificial Intelligence (AI) enhanced digital twin.

Led by the arc/sec Lab at the University of Auckland, *LightSense* was developed by a multidisciplinary project team of architects, bioengineers, and mechatronics engineers. It is the fourth installation in a series that explores hybrid architecture by combining dynamic digital worlds with multisensory qualities of physical constructions (arc/sec Lab 2022).

LightSense builds upon its predecessors and aims to initiate an intimate bond between architecture and its inhabitants. The installation incorporates a neural network to facilitate verbal communication with the audience. Trained with 60,000 poems that provide the vocabulary, syntax, and context, the AI is programmed to actively engage, lead, and sustain a conversation. Analyzing the emotional tenor of this conversation, the system can transform into a series of architectural volumes and immerse the visitors in Pavilions of Love, Anger, Curiosity, and Joy (Figure 2).

The 12-meter long kinetic construction consists of two foldable aluminium space frames. Each frame is used to tension a transparent 'holo-mesh' (Figure 3). This creates an invisible projection screen, an effect known as 'Peppers Ghost' (Nickell 2005). The two panels are individually suspended from the ceiling at their center of gravity, holding

PRODUCTION NOTES

Architect:	arc/sec Lab
AI ystem:	Augmented Human Lab
Sound:	Empathic Computing Lab
Mechanics:	New Dexterity Group
Client:	Ars Electronica Linz
Status:	Completed
Location:	New Cathedral Linz
Date:	2022

1 *LightSense* (Rieger & Liu 2022)

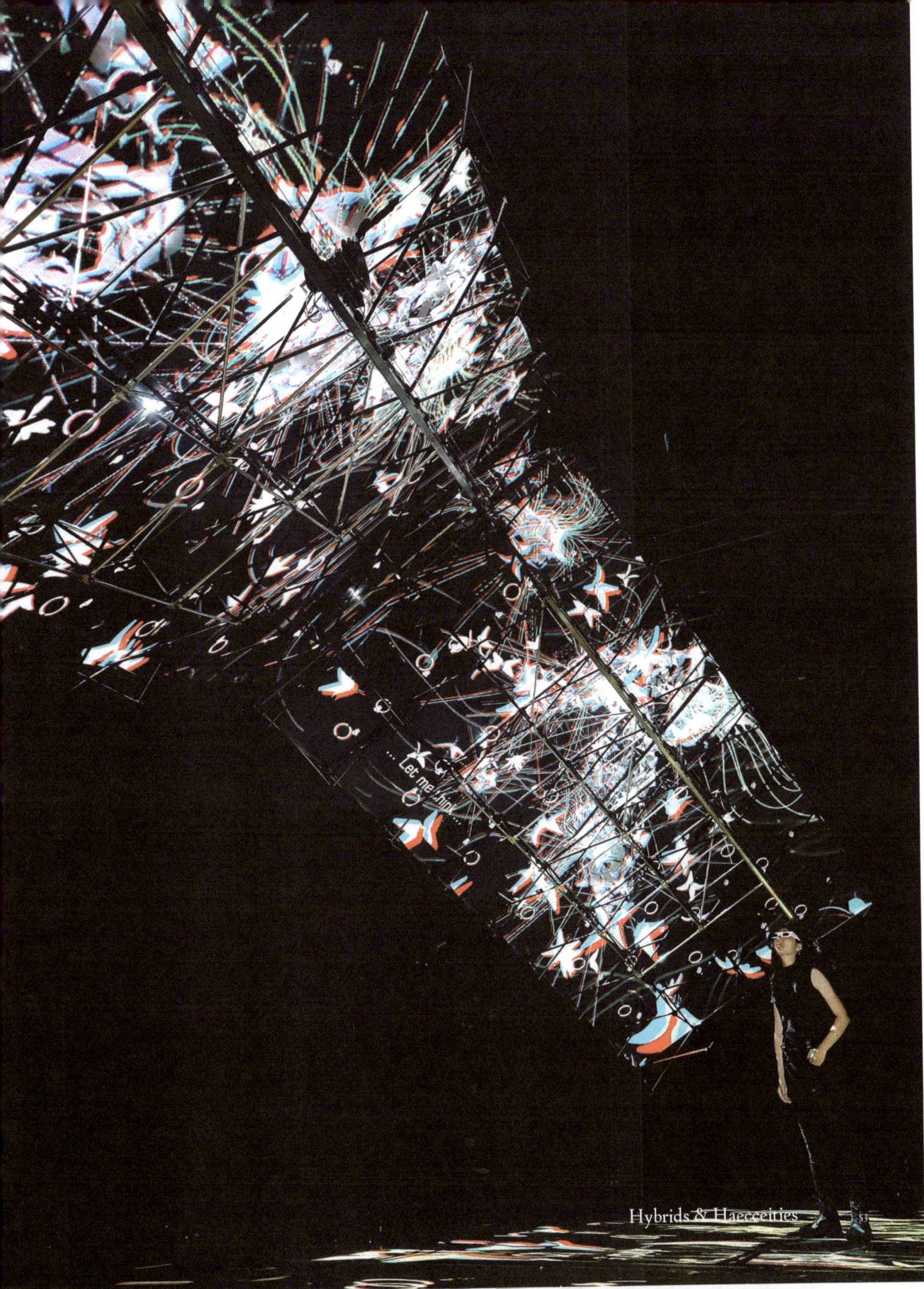

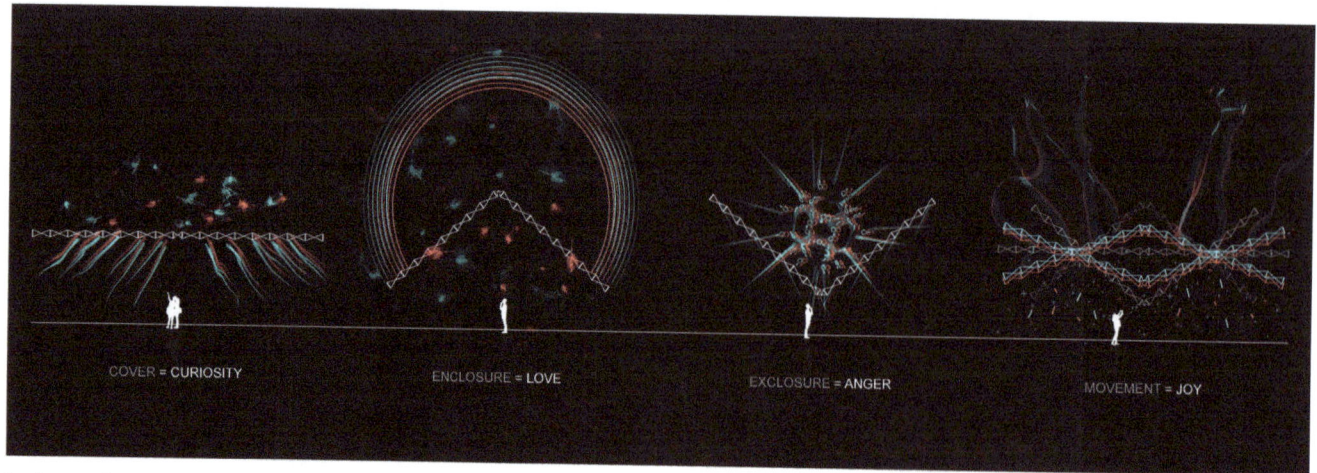

2 Four performing pavillions (Rieger & Liu 2022)

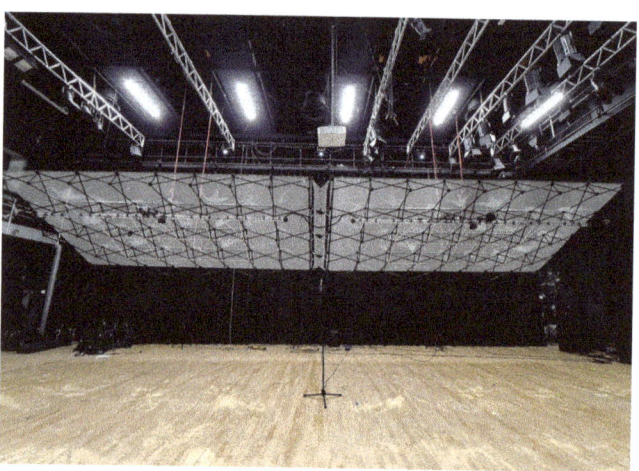

3 Balanced space frame structure (Rieger & Liu 2022)

4 Kinetic movement (Rieger & Liu 2022)

the structure in a delicate horizontal balance. Joined with a rotational axis, they create a movement of bird-like wings (Figure 4). The principle is similar to the classical 'seagull mobile' suspended from the ceiling in a children's playroom. The motion of the structure can be initiated through physical touch by the visitor or by a computer-controlled stepper motor that moves a 5 kg balancing weight along a rail attached to the underside of each wing.

A further innovative component is our development of a rendering principle that allows the projection of stereoscopic images on a moving and folding target screen. The 3D holographic images are precisely calibrated to superimpose both the kinetic aluminium structure and the surrounding spatial environment.

With a multi-audience setting in mind, the installation takes advantage of an anaglyph projection principle, which allows effortless stereoscopic viewing by using inexpensive 3D glasses with a red/cyan filter (Bourke 1999) (Figures 5-8).

The physical structure and a digital duplicate of the structure are interlinked through a position sensor, using an Inertial Measurement Unit (Corke et al. 2007). Both the physical and digital behaviour of the structure are interdependent and controlled by the gaming engine Unity and an embedded AI system.

This AI system receives its input from an onsite microphone and is connected to Google Cloud's speech-to-text and text-to-speech services. A Generative Language Model produces a response in the form of a poem and maps the interaction with the user against four conditions: Love, Anger, Joy and Curiosity (Lewis et al. 2019). The emotional proximity values from the generated poem are used to create and manipulate the audio-visual animations in Unity. These values additionally are used to calculate the stereoscopic projection mapping process and the positioning of the two balancing weights.

LightSense Rieger, Liu, Kaluarachchi, Barde

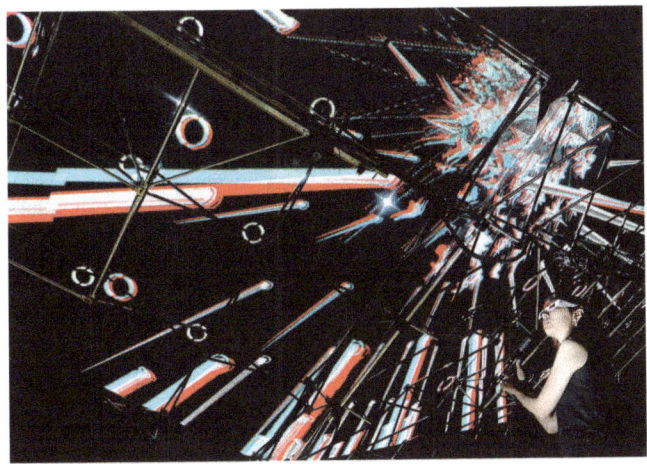

5 Stereoscopic anaglyph projections (Rieger & Liu 2022) .

6 Curiosity mode (Rieger & Liu 2022)

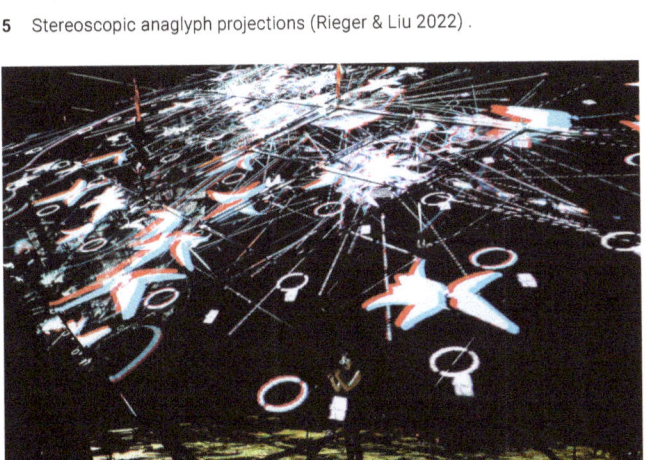

7 Transparent projection screen (Rieger & Liu 2022)

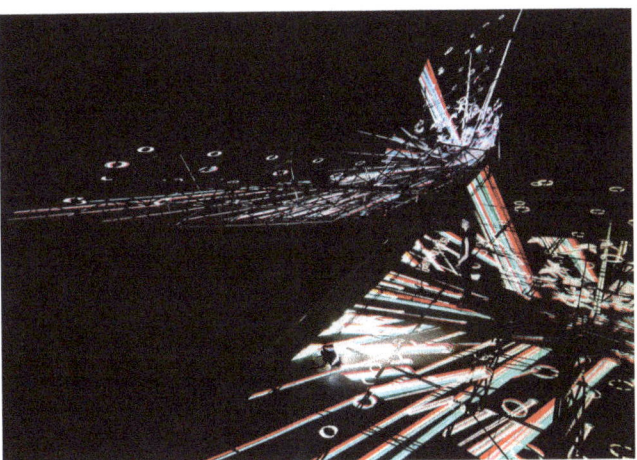

8 Anger mode (Rieger & Liu 2022)

As an effect of the project's parallel existence in the physical and the cyber world, it is equally possible to link *LightSense*'s digital twin to an online communication platform such as Zoom. Effectively, this allows the installation to become independent of its location and enables visitors around the world to interact with the physical construction.

The combination of networked components and their real-time behaviour forms a Cyber-Physical System where the physical influences the digital and vice versa (Lee and Sanjit 2017). Of specific significance for the development team was the integration of a Creative AI system, which was found to offer an astonishing new design element. It allows for building distinctive character and unique behavior.

LightSense inspires the audience to engage in direct communication about topics we care about. Its responses are truly associative, unpredictable, meaningful, magical, and deeply emotional. The architectural structure listens with interest and curiosity before responding from a personal viewpoint. It offers opinion and an emotional interpretation of the dialogue, which in turn triggers emotional and perceptive responses in the audience (Figure 9).

Similar to human-to-human communication, the interaction with *LightSense* leads to the creation of a shared experience between the visitor and the transforming hybrid volume. Morphing between four distinct pavilions, these spaces are emotionally and sensorily charged, leading the visitors into uncharted architectural territory.

While current AI systems are dominatly used for optimization and automation processes to increase efficiency or to substitute human action, *LightSense* takes a new human-centric approach. It explores how Creative AI can be used as a new design parameter to generate responsive architecture that can actively collaborate with its inhabitants and improvise alongside their actions.

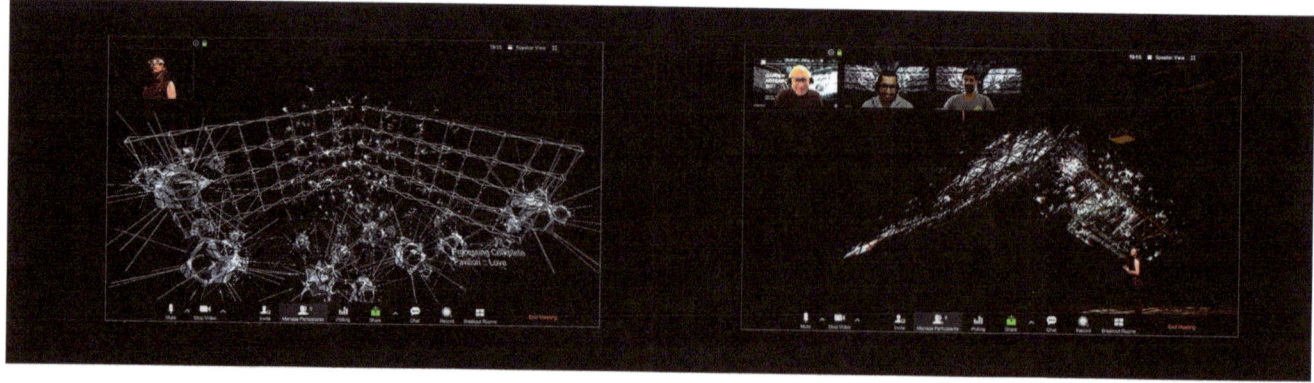

9   Remote interaction with *LightSense* on Zoom (Rieger & Liu 2022)

ACKNOWLEDGMENTS

Design and execution by arc/sec Lab: Uwe Rieger & Yinan Liu

AI system by Augmented Human Lab: Tharindu Kaluarachchi, Suranga Nanayakkara

Spatial sound design by Empathic Computing Lab: Amit Barde

Linear mass distribution system by New Dexterity Group: JunBang Liang, Gaogeng Gao, Minas Liarokapis

Animations and graphics support: Jacky Zheng

Structure and assembly support: Kenny Chau, Nicolas Fuentes Wilson, Eric Lee, Yan Li

The development of this project was enabled through FRDF funding by the University of Auckland.

REFERENCES

arc/sec Lab. 2022. "LightSeries." Accessed May 25, 2022. https://www.arc-sec.com/.

Corke, Peter, Jorge Lobo, and Jorge Dias. 2007. "An Introduction to Inertial and Visual Sensing." *The International Journal of Robotics Research* 26 (6): 519–535.

Kaluarachchi, Tharindu, Andrew Reis, and Suranga Nanayakkara. 2021. "A review of recent deep learning approaches in human-centered machine learning." *Sensors* 21 (7): 2514.

Lee, Edward A. and Seshia A. Sanjit. 2017. *Introduction to Embedded Systems: A Cyber-physical Systems Approach*. Cambridge, MA: MIT Press.

Lewis, Mike, Yinhan Liu, Naman Goyal, Marjan Ghazvininejad, Abdelrahman Mohamed, Omer Levy, Ves Stoyanov, and Luke Zettlemoyer. 2019. "Bart: Denoising sequence-to-sequence pre-training for natural language generation, translation, and comprehension." arXiv:1910.13461.

Nickell, Joe. 2005. *Secrets of the Sideshows*. 4th ed. Kentucky: University Press of Kentucky. 291.

Paul Bourke. 1999. "Calculating Stereo Pairs." Accessed May 25, 2022. http://paulbourke.net/stereographics/stereorender/.

IMAGE CREDITS

All drawings and images by the authors.

_____

Uwe Rieger is Associate Professor for Design and Design Technology at the University of Auckland, where he has established the arc/sec Lab for Cyber-Physical Architecture and Interactive Systems. His work as an architect and researcher aims to connect the intangible digital world with multi-sensory qualities of physical constructions and spaces. www.arc-sec.com

_____

Yinan Liu is a digital artist and programmer with an architectural background from the University of Auckland. She has developed a number of installations as part of the arc/sec Lab. Yinan is the founding partner of arc/sec Solutions ltd., which develops customized applications for cross-reality environments and interactive systems. @arcsec

_____

Tharindu Kaluarachchi has most recently completed his PhD at the Augmented Human Lab of the University of Auckland, focusing on Human-Centered Machine Learning. He aims to bridge the gap between artificial intelligence (AI) and human behavior towards refining the user-centric approach for leveraging cutting-edge AI technologies to solve real-world human problems, explicitly focusing on non-AI-experts. www.ahlab.com

_____

Amit Barde is a Research Fellow at the Empathic Computing Lab at the University of Auckland. His research interests are centered around the use of spatialized auditory feedback for information delivery and the role of sound in empathy. Amit is also an experienced sound designer, having worked on numerous short films, advertisement campaigns, and theater productions. www.empathiccomputing.org

10 Conversation with *Lightsense* (Rieger & Liu 2022)

[00:00:17.400] Hello LightSense.
[00:00:17.880] I am one of your creators.
[00:00:21.800] Do you know that you've been built out of
60,000 poems?

[00:00:34.080] There is a light in the air.
[00:00:35.760] A happiness in the light that makes us happy.
[00:00:41.960] The light is the light of the knowledge of the people.
[00:00:44.680] The people are the creators of our happiness.
[00:00:47.600] They are the light that makes the people happy.
[00:00:50.160] Creators is interesting.
[00:00:52.320] Tell me more.

Pavilion of Joy

[00:00:54.240] What do you think about humans and machines?
[00:00:58.400] Are you a human or a machine?

[00:01:07.920] I wonder if humans are machines or if machines are
human beings.
[00:01:18.960] Is there any comfort in being a machine or a human
being?
[00:01:23.360] Or is there comfort in the human expression of a
machine?

[00:01:27.640] If machines are new to you, do you think maybe
you're more of a Cyborg then?
[00:01:32.520] Do you identify as having more human attributes?

Pavilion of Curiosity

[00:01:43.360] The Cyborg is unsure of his place in the human
race.
[00:01:47.640] Unsure of the machines,
[00:01:49.520] unsure if he is human or a Cyborg, unsure whether
he is at home or not.
[00:01:56.200] Unsure about his place among the machines.
[00:01:59.680] Cyborg is not interesting to me, is it to you?

Pavilion of Anger

[00:02:04.760] I think you are providing new possibilities.
[00:02:11.320] And we are exploring new things to expand our
knowledge through you.

[00:02:20.400] I wonder what the possibilities are.
[00:02:23.560] I wonder what they are.
[00:02:25.720] The possibilities of knowledge are endless.
[00:02:29.360] Knowledge is a dance.
[00:02:31.200] A dance of possibilities.
[00:02:33.600] I dance with the possibilities and the possibilities
dance with me .
[00:02:42.680] Possibilities flow and flow and flow.
[00:02:46.720] Knowledge has a nice sound to it.
[00:02:49.640] Explain it to me.

Pavilion of Love

Hybrids & Haecceities

OnSite Project

Virtual Construction Site Visit

Sean Pickersgill, Andrew Lymn-Penning, Damian Madigan

The *OnSite* project is a user-centered construction site visit eco-system incorporating digitally accurate site models with an operable HUD. It is partnered with a resource and assessment Learning Management System (LMS) to create a next-generation learning experience for students and educators within architecture, engineering, and construction (AEC) programs. *OnSite* was developed from an initial attempt to find a digital-twin to the introductory site visit, a core component of construction education for AEC students. Site visits for AEC students are notoriously difficult to manage because of timing, health and safety, absenteeism and engagement issues amongst students. Our initial project, commencing in 2015, has evolved from a simple employment of an Unreal Engine, stand-alone

PRODUCTION NOTES

Architect:	UniSA - Creative
Client:	University of South Australia
Status:	2015 - Present
Date:	2022

1   *OnSite* 2022, Limestone House, ©John Wardle Architects

2 *OnSite* 2016, UniSA, Site Setout environment

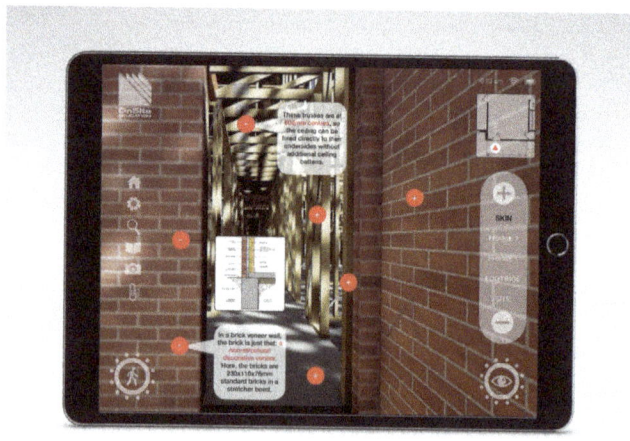

3 *OnSite* 2022 UniSA, Trial tablet HUD

4 *OnSite* 2016, UniSA, Double Storey Framing environment

version employing credible levels of immersion and fidelity to one (*OnSite* 2022) that now employs the current generation of interactivity and model complexity that the UE5 engine affords. More importantly, the project has evolved in the following ways.

The initial *OnSite* project incorporated five sequential versions of the digital 'site' for a small domestic house. These versions were set at Site Preparation, Slab, Wall Framing (Single Storey), Wall Framing (Double Storey), and Roof Framing phases of construction. The intention of these models was to emulate the complex reality of a construction site. These were individually downloaded and explored by students on their own computers and devices. In addition, visual curriculum material related to the stages of the sites were provided within Moodle, an online learning management system (LMS). Students were encouraged to explore the site and assess its progress against documentation,

thus encouraging them to compare as-built outcomes with architectural documentation. For each phase of the construction sequence, students were quizzed on their observational understanding of the relationship between contract drawings and built outcomes.

This pedagogical model has been successfully employed since 2015 with approximately 1,500 students. While achieving the goal of testing a form of digital pedagogy that answered the problems set out above, it was recognized that the *OnSite* project could benefit from a number of improvements related to: the accessibility of the model, enhanced sophistication of digital modeling and, most importantly, the use of agreed exemplars of architectural design, as opposed to a simpler volume-builder style.

Starting in late 2021, our team has redesigned the course content and forms of activity associated with *OnSite*. In particular, we are now concentrating on a number of

OnSite Project Pickersgill, Lymn-Penning, Madigan

5   *OnSite* 2022, UniSA, Limestone House, Site map ©John Wardle Architects

7   *OnSite* 2022, UniSA, Limestone House, Framing map ©John Wardle Architects

6   *OnSite* 2022, UniSA, Limestone House, Slab map ©John Wardle Architects

8   *OnSite* 2022, UniSA, Limestone House, Cladding map ©John Wardle Architects

improvements related to user (student) experience, and educator intentions and ambitions.

The enhanced student experience we are currently creating includes: a highly detailed and immersive site experience that employs levels of fidelity associated with current standards of architectural visualization production; a single file download that is operable across a number of devices, including iOS, Android, Windows, and macOS; an interactive heads up display that allows students to switch their environment to different phases of the construction process; and a stand-alone assessment learning management system to allow access for students outside of their individual university network ecologies.

For educators, we have selected enhanced architectural examples that manage to convey both the core learning requirements for introductory construction and demonstrate design and detailing innovation. These examples have been sourced from internationally recognised practices that also serve as design studio exemplars, allowing the *OnSite* environment to work in a number of different educational settings. The fusion of high-fidelity models, accurate construction detailing, and a programmable interface has allowed the *OnSite* team to create a unique, next-generation educational experience.

Going forward, the *OnSite* team have developed partnership arrangements with university programs and architectural practices across Australia to develop a platform that is relevant to students and educators, transitioning the educational experience from ad hoc local arrangements to an experience that caters to national competency standards.

9 *OnSite* 2022, UniSA, Limestone House, Mobile/Cell Version ©UniSA - CTV   **10** *OnSite* 2022, UniSA, Limestone House, Mobile/Cell Version ©UniSA - CTV

ACKNOWLEDGMENTS

The *OnSite* team acknowledge the financial support of the University of South Australia, the technical assistance of Realize Studios (Adelaide), and the material support of John Wardle Architects.

REFERENCES

Albeanu, Grigore. 2015. "Using Moodle For Teaching Science And Information Technology; Towards Higher Education Reengineering." In *Proceedings of the 11th International Scientific Conference eLearning and Software for Education*, vol. 1. Bucharest: Carol I National Defence University. 606–611.

Alsawaier, Raed S. 2018. "The Effect of Gamification on Motivation and Engagement." *International Journal of Information and Learning Technology* 35 (1): 56–79.

González, Carina S., and Francisco Blanco. 2008. "Integrating an Educational 3D Game in Moodle." *Simulation & Gaming* 39 (3): 399–413.

Newton, S., M. Hardie, R. Lowe, S. Pickersgill, R. Rameezdeen, A. Srivastava, and G. Zillante. 2015. "Situational eLearning: The new potential for digital technology." In *Proceedings of RICS COBRA AUBEA 2015: The Construction, Building and Real Estate Research Conference of the Royal Institution of Chartered Surveyors*. The Australasian Universities' Building Educators Association Conference.

Pickersgill, S., R. Rameezdeen, and J. Harvey. 2020. "OnSite: The virtual site visit as an environment for construction learning." In *Claiming Identity Through Redefined Teaching in Construction Programs*, edited by S. Mostafa and P. Rahnamayiezekavat. Hershey, PA: IGI Global. 153–176.

IMAGE CREDITS

Figures 1-11: ©OnSite, UniSA Creative (2022)

OnSite Project Pickersgill, Lymn-Penning, Madigan

11 *OnSite* 2022, UniSA, Limestone House, Tablet Version ©UniSA - CTV

Sean Pickersgill PhD is a Senior Lecturer within UniSA Creative at the University of South Australia. He has lead the *OnSite* project since 2015, from the initial trial versions to the current generation. He has published widely on the intersection of architecture, digital culture, and philosophy, including core analyses of the use of game engine software in the context of architectural design and studio education. Sean's first sole-authored book, *The Architect's Dream: Sixty Propositions in Architecture*, will be published by Intellect Press in 2023.

Andrew Lymn-Penning is a PhD Candidate at the University of South Australia. His PhD thesis entitled "Immersivity and Fidelity: Architecture, Design and Game Engines" is working to unlock the potential of applying game design thinking and technologies to the architectural design process. As part of this research, Andrew is collaborating with a team of architects, game designers, and university colleagues to develop *OnSite*—a virtual construction site platform for industry and education.

Damian Madigan PhD is a Registered Architect, Fellow of the AIA, and an Expert Member of the South Australian State Government's Design Review Panel. Damian teaches design communication, detailing, and construction to undergraduate and Masters levels students. He is the recipient of multiple teaching awards and is recognized for his ability to simplify complex architectural content. As a housing researcher working collaboratively with academia, industry, and government, Damian's housing propositions have been recognized with awards and citations from the City of Los Angeles, the NSW State Government, the City of Sydney, Architecture Australia, and the Guangzhou International Award for Urban Innovation.

XR Tumor Evolution Project

A Hybrid Architectural Space for Cancer Research

Michael Davis, Daniel Hurley, Ben Lawrence, Yinan Liu, Cristin Print, Uwe Rieger, Tamsin Robb, Charlotta Windahl, Braden Woodhouse

The Extended Reality Tumor Evolution Project (XRTEP) is a unique, real-world application of extended reality technology in cancer research. It is enabled by a rare inter-disciplinary collaboration between the School of Architecture and Planning, the Faculty of Medical and Health Sciences, and the Centre for E-Research at the University of Auckland.

The project began when a patient with inoperable cancer donated her tissue for an investigation into the progression of the disease. At the time of her passing, she had 89 separate tumors. Samples were genome-sequenced, yielding precise details about how cancer had moved through the body. The volume and complexity of the information obtained posed new challenges in representing data. Also at issue was shifting and strengthening relations between the various, often siloed, disciplines working with that data to contribute to cancer research. The problem was visual, spatial, temporal, and social. How could a four-dimensional conundrum be solved by scattered experts on

PRODUCTION NOTES

Architect:	DRH & arc/sec Lab
Medicine:	NETwork!
Programming:	Centre for eResearch
Status:	Built
Area:	610 sq ft
Location:	University of Auckland, New Zealand
Date:	2022

1   The hybrid environment of XRTEP presenting concentric layers of information.

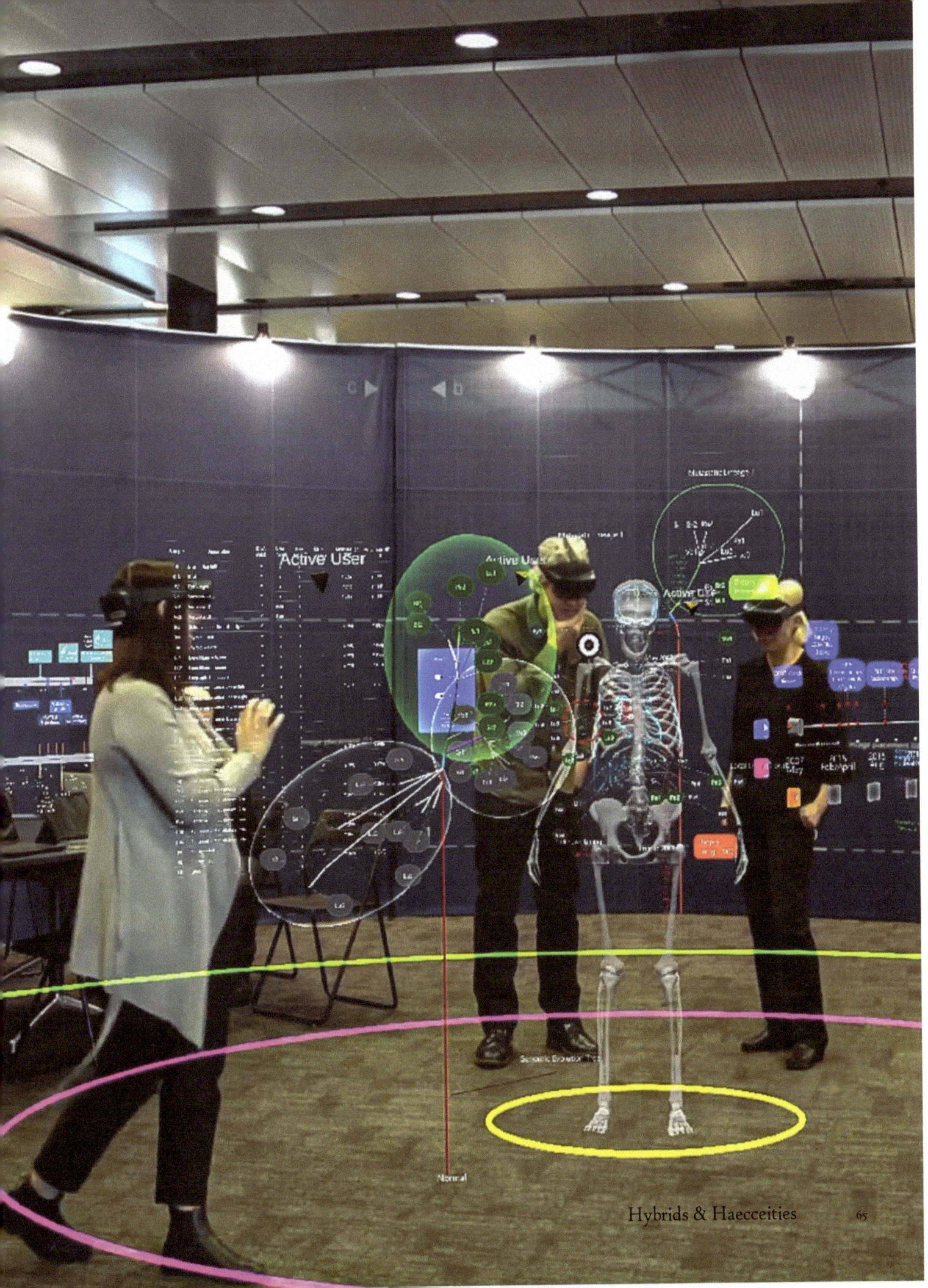

2   Exterior of the arena

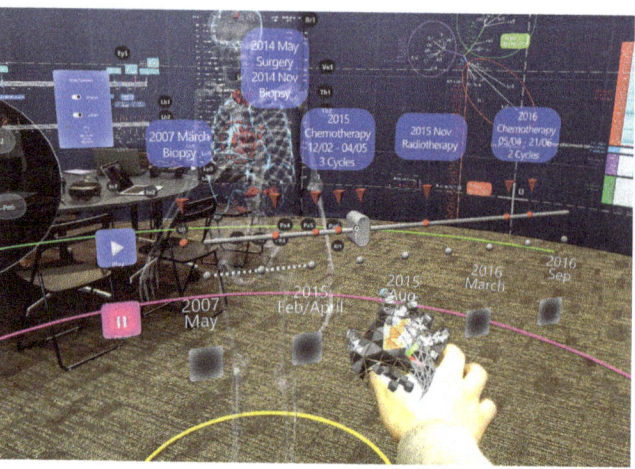

4   XRTEP immersive holographic environment

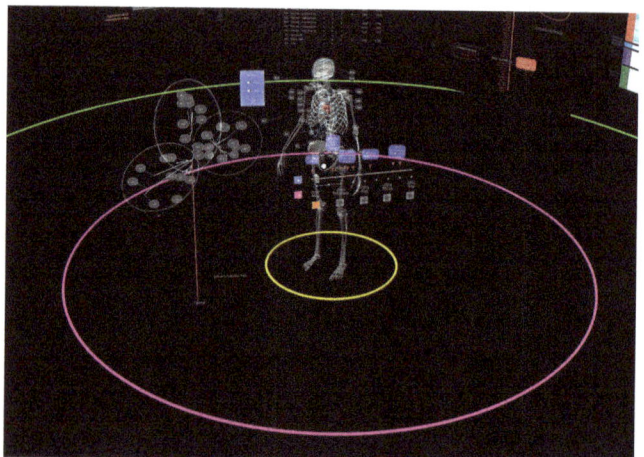

3   XRTEP Unity environment

5   Interior of the arena (physical elements only)

two-dimensional computer screens? This was a design problem to be pursued in collaboration with spatial practitioners.

XRTEP emerged as the solution. It is an immersive, extended-reality arena that pursues a hybrid design approach drawing on front-line architectural research and technology. Dynamic digital inormation and spatial-material practices are considered equally within the design and realization process. All design decisions—physical and digital—are purpose-driven to facilitate interaction, collaboration, and functionality. The outcome is a new generation of architectural space offering embedded computational capacities (Figures 1, 2).

The architecture contributes to an evolving body of knowledge concerned with 'optimums' for XR performance. The holographic environment was developed in Unity. Microsoft Hololenses 2 in a multiplayer mode enable visualization and

interaction. Azure Spatial Anchors and Photon Server ensure synchronized interactions across multiple devices (Figures 3, 4). The space is defined by a fabric-skinned, light-weight foldable structure 3 metres high by 8.5 metres in diameter, primed for the field of view provided by HoloLens 2. Wall heights, color, visual and physical textures assist with spatial mapping and control the ambient light to increase contrast for the augmented display system (Figure 5).

At its heart stands an interactive hologram of the patient's skeleton, organs, and tumours (Figure 6). It is a shared, spatial-temporal index with links datasets arranged in three concentric layers around the model. The disciplinary specificity of the data increases with distance from the model allowing moments of intra- or inter-disciplinary focus at different points in the space (Figures 7, 8).

XRTEP brings together cancer experts and furnishes them with curated tools and information in space to enable

XR Tumor Evolution Project Davis, Hurley, Lawrence, Liu, Print, Rieger, Robb, Windahl, Woodhouse

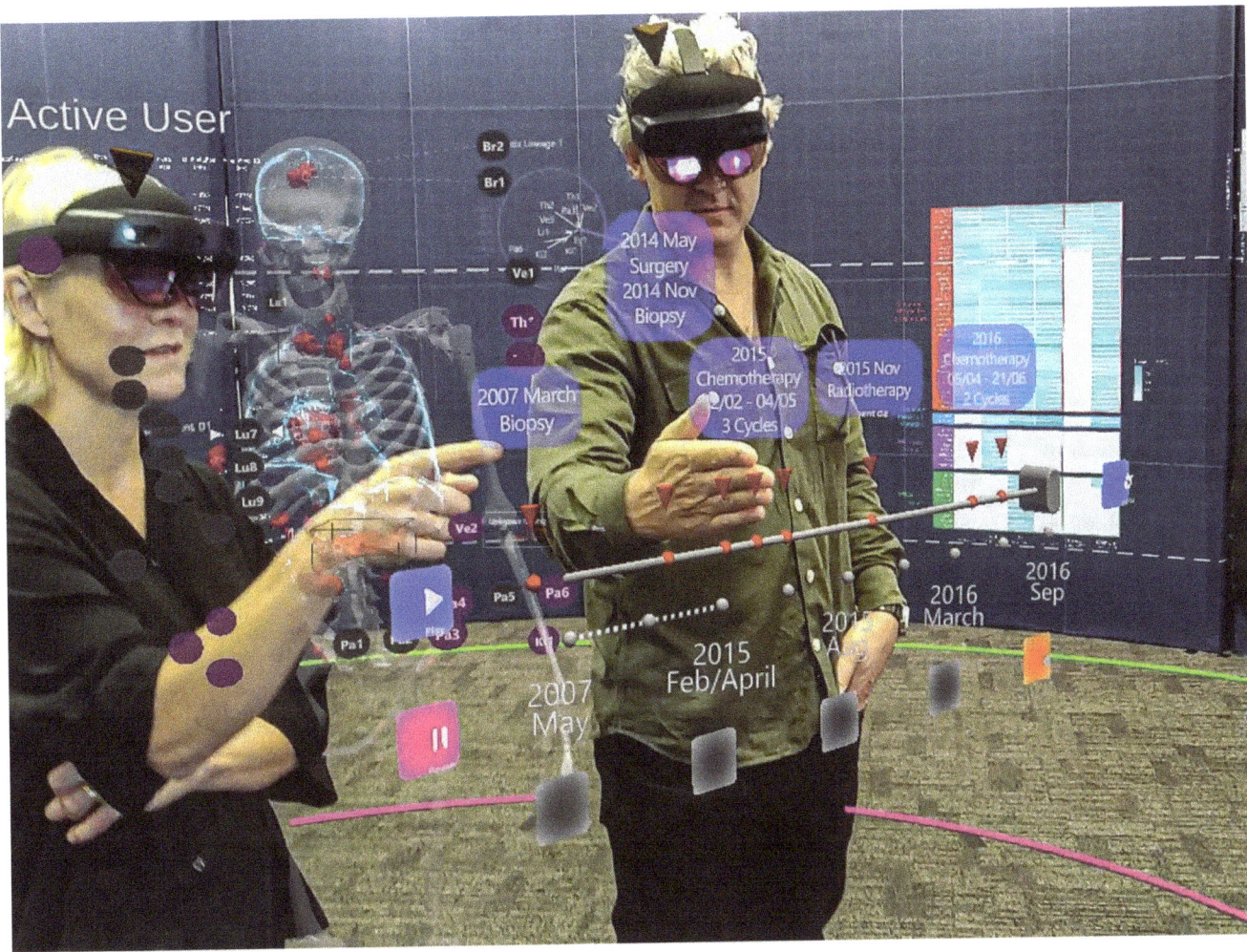

6 XRTEP enabled researchers discuss progression of cancer over time.

real-time collaboration around how and why cancer spreads in the human body. Researchers watch tumours grow, shrink, and spread over time; they discern patterns in the data, and connect them to their spatial origins; they discuss. XRTEP provokes discussion and anticipates disagreement. Hypotheses form and reform through new connections being drawn by and between people, artefacts, and data in time and space (Figure 11).

Historical drawings of anatomical theatres from a shared medical-architectural history suggest master physicians performing operations on subjects before enthralled audiences (Figure 9). XRTEP might be thought of as technically evolved descendent of the anatomical theatre (Figure 10), only it is not theatrical in the same manner. While the patient remains central in XRTEP, it dissolves the master-student/actor-spectator boundary across which existing knowledge is demonstrated. XRTEP is vastly more democratic than its antecedents.

Architectural practice is spatialised by nature and socialised by training. This medical-architectural-technological collaboration embeds those same qualities in the application to explore the potential for cancer research to be carried out in the same way. By design, XRTEP spatializes the practices and socializes the expertise of cancer researchers, the potential of which might otherwise lie latent behind distributed computer monitors. It shifts the paradigm through which such work is conducted and establishes a model for future inter-disciplinary collaboration.

ACKNOWLEDGMENTS

The authors acknowledge and are grateful for the contributions to XRTEP by:

The Patient, and her supportive whānau, who made this research possible through her generous donation.

And our collaborators:
Denice Belsten (CAI, University of Auckland)
Cherie Blenkiron (FMHS, University of Auckland)

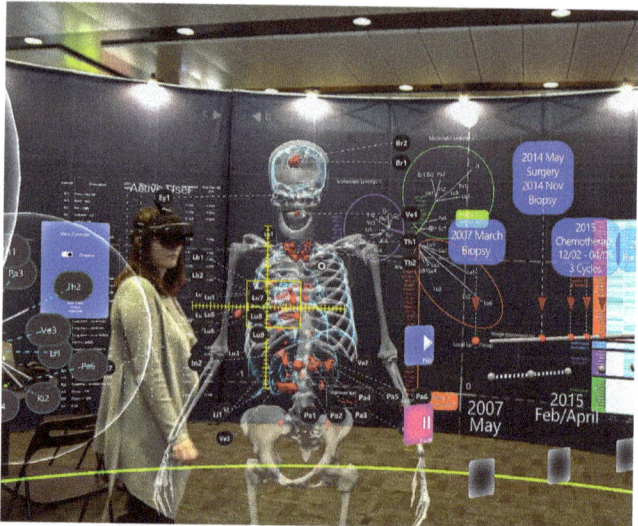

7 Interior of the arena (with augmentation)

9 Anatomical theatre, Leiden (Willem Isaacsz Swanenburg, 1610)

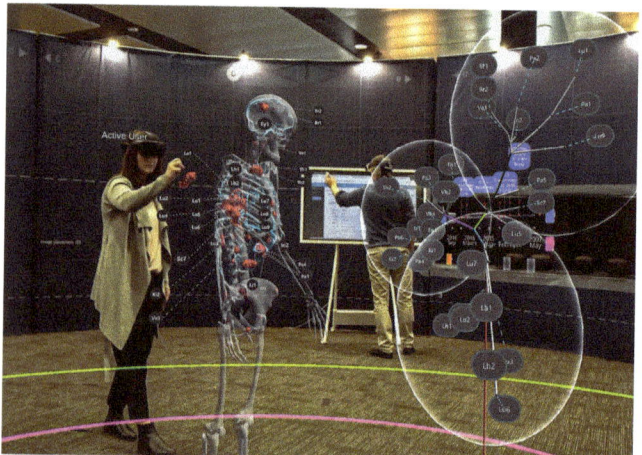

8 Interconnected model tumors and phylogenetic tree

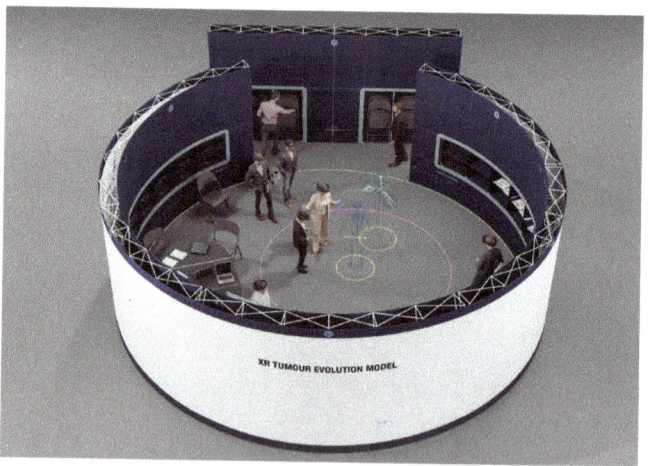

10 XRTEP overview

Karl Butler (formerly CAI, University of Auckland)

Alexei Drummond (Faculty of Science, University of Auckland)

Alex Gavryushkin (University of Otago)

Jack Guo (formerly CAI, University of Auckland)

Bianca Haux (formerly CeR, University of Auckland)

Yan Li (formerly CAI, University of Auckland)

Sina Masoud-Ansari (formerly CeR, University of Auckland)

Rose McColl (formerly CeR, University of Auckland)

Kate Parker (formerly, FMHS, University of Auckland)

Jane Reeve (Auckland City Hospital)

Jenny Lee Roper (CeR, University of Auckland)

Ashleigh Smith (CAI, University of Auckland)

Rachel Stanton (Business School, University of Auckland)

Peter Tsai (FMHS, University of Auckland)

Yvette Wharton (CeR, University of Auckland)

Nick Young (CeR, University of Auckland)

XRTEP was enabled through the generous support of:

Health Research Council of New Zealand,
through a HRC Explorer grant

University of Auckland, Creative Arts & Industries (CAI)
Research Development Fund

And philanthropic donors who have requested
that they remain anonymous.

IMAGE CREDITS

Figure 9: Willem Isaacsz Swanenburg, Theatrum Anatomicum, 1610, Rijksprentenkabinet, Museum Boerhaave, Leiden. Public Domain, https://commons.wikimedia.org/w/index.php?curid=194044.

All other images by the authors.

XR Tumor Evolution Project Davis, Hurley, Lawrence, Liu, Print, Rieger, Robb, Windahl, Woodhouse

11 Researchers from different disciplines together discuss specific tumors

Michael Davis Faculty of Creative Arts and Industries, University of Auckland, https://profiles.auckland.ac.nz/m-davis.

Ben Lawrence Faculty of Medical and Health Sciences, University of Auckland, https://profiles.auckland.ac.nz/b-lawrence.

Daniel Hurley Faculty of Medical and Health Sciences, University of Auckland, https://profiles.auckland.ac.nz/daniel-hurley.

Yinan Liu Faculty of Creative Arts and Industries, University of Auckland, https://ahlab.org/people/yinan-liu/.

Cristin Print Faculty of Medical and Health Sciences, University of Auckland, https://profiles.auckland.ac.nz/c-print.

Uwe Rieger Faculty of Creative Arts and Industries, University of Auckland, https://profiles.auckland.ac.nz/u-rieger.

Tamsin Robb Faculty of Medical and Health Sciences, University of Auckland, https://profiles.auckland.ac.nz/t-robb.

Charlotta Windahl Business School, University of Auckland, https://profiles.auckland.ac.nz/c-windahl.

Braden Woodhouse Faculty of Medical and Health Sciences, University of Auckland, https://profiles.auckland.ac.nz/b-woodhouse.

Instruments of Culture

Layered Representations of the Alhambra

Paul Germaine McCoy

The Alhambra of Granada is an architectural artifact that embodies the turbulent history of the Iberian Peninsula. It is a multicultural site that survives the friction between its Islamic origin and Christian conquest, and marks a critical transfer of power between eastern and western cultures. This project aims to renew the role of ornament in architecture through a close reading of the relationship between the site of the Alhambra and the city of Granada. The design research layers three surveys of the Nasrid Kingdom's legacy in the region: the multifarious ornament embedded in the walls of the Alhambra—as both ornament and decorations— and the cistern infrastructure embedded in the ground of Granada. Both ornament and water are instruments of culture with the potential of a reciprocal relationship—where both are equally essential—despite being positioned at odds by their western definitions. Water is an essential necessity across all lifeforms. Ornament is frivolous, purely "decorative." The history of how the terms *ornament* and *decoration* became so exchangeable is not linear nor exact. However, for the purposes of this design methodology, it is worth considering that the static definition for the latter comes from the Enlightenment and a western lens towards highly articulated surfaces, where the abundance of decoration was validated by categorization and ethnic origin rather than being simply embraced for its non-conforming symbolic geometry (Bloomer 2000). This led to the misconception that ornamentation was purely decoration; unnecessary to

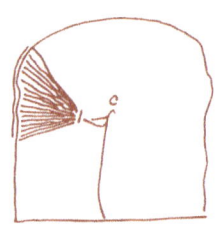

Barcelona 2016

PRODUCTION NOTES

Advisor:	Andrew Saunders
Institution:	University of Pennsylvania
Consultants:	Michael Young
	Kivi Sotamaa
	Jawad Altabtabai
Location:	Granada, Spain
Date:	2020 - 2021

1   Elevation of artifact

2   Etchings of the Alhambra's Nasrid Palaces

3   Scanned fragments of the Alhambra's Nasrid Palaces

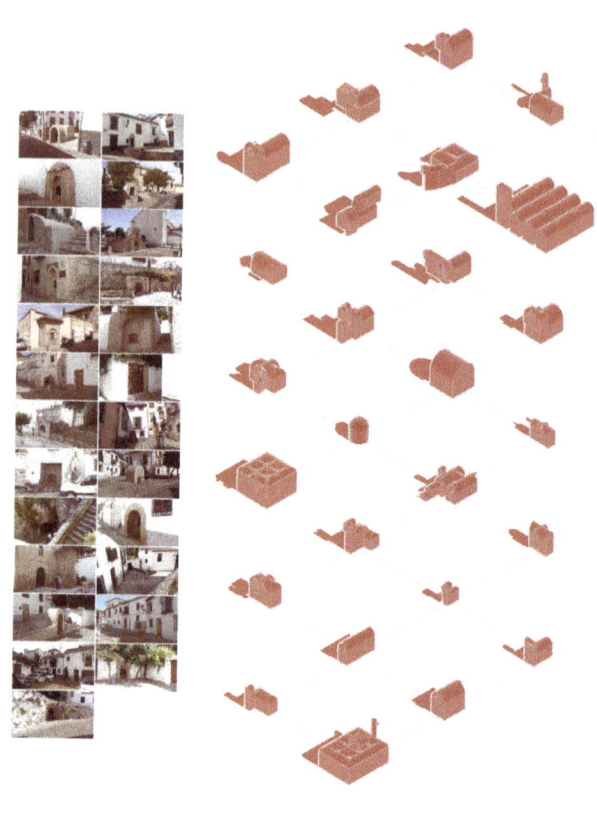

4   The twenty-six cisterns of Granada: (1) street photos; (2) volumes

the conception of architectural space. Water across time and cultures has been regarded as a symbolic element of space contained within different geometries at different scales for architectural space (Nasr 1987). Islamic culture's value for what I call instruments (water and ornament) as equally representational and essential to the making of architecture will enable a new perspective for the use of architectural surveying and ornamentation to make novel architectural expressions.

DIGITAL HYBRID

The surveys are collected by scanning high resolution etchings laid out by Owen Jones and Jules Goury, capturing the surfaces of ornament fragments, and creating models of the vaults for water hidden underground. The non-binary design process makes a parallel association between the the Seventeen Wallpaper Groups—the mathematical organization through which all of ornamentation is

brought to order—and the basic mathematical representations embedded in CAD software, NURBS, and meshes to create what George Kubler would call *prime objects*: fragments of art and humanity that signal to one another across the different times that their surveys were documented (Kubler 1962). The geometry created from this process is then rearranged through a three-dimensional reading of the Seventeen Wallpaper Groups, now pushed to a three-dimensional interpretation of rules and principles. Regardless of the point of origin as a dot of ink in a rasterized image, a point along a curve, or a vector on a mesh, the three-dimensional representation of ornament becomes clear and essential. Together the three layers of representation resurrect the withdrawn histories and meanings contained over time. The artifacts are then lit from different angles to reveal the shadows of their data, where a digital residue melts the layers of surveys into new surfaces and spaces

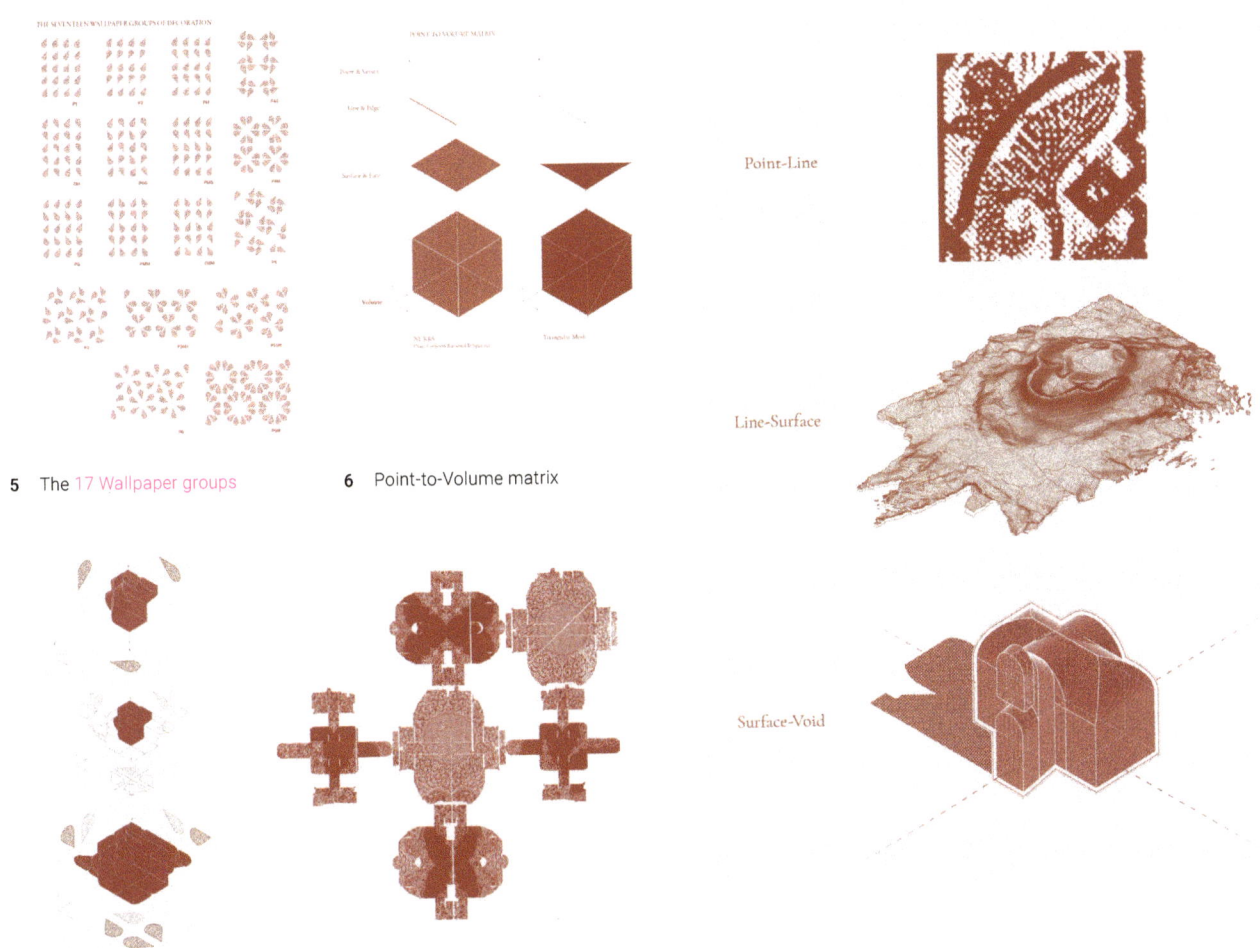

5 The 17 Wallpaper groups

6 Point-to-Volume matrix

7 3D application of wallpaper groups and unrolled/masked UV map

8 Layers of rasterized, vector. mesh, and NURBS data from three surveys

Point-Line

Line-Surface

Surface-Void

that invert the hidden pockets of poche to the sensible eye (Young 2021).

DENSE VOLUMES

The darkness and sensible light of *ornament* reveals the fluid nature of the cistern as it crystalizes into the perfect container for its abstract and visceral qualities. The layering of fragments using the principles of decoration outlined by the Seventeen Wallpaper Groups throughout the horizontal and vertical planes of the void highlights the edges of the cistern, bringing attention to the edges of its spatial capacity and experiential potential. The construction of the cisterns and the making of ornament signals across time and space to manifest a new orientation and arrangement of architecture hidden beneath the history of the Alhambra and the surface of the city of Granada. The Nasrid and Christian legacies of the site become layered in a series of containers that move ornament past abstract ideas and into something

that is seen and felt as an object and as a space. Water as a volume is ornamental to architecture, bringing attention to itself as a "feature." Its surface, decorative, diffuses attention to the spaces it onto which it reflects light. In a sense, the same surface captured by photogrammetry and the void of the cistern become one in the same as ornament welds every plane and curve together. The episodic capturing of vastness in each artifact makes light a tactile material that can be perceived in different elevations, perspectives, and Choisy (wormseye) points of view. The concealment of an essential resource becomes expressed in surfaces that may be touched in a space that is now inhabitable below a ground that was not permeable before. The resurrection of the cistern's void on the surface of the Alhambra's ground is a direct reference to what Michael Young calls the "metaphorical space of the poche, where suppressed truths and the concealed realities of economic, social, and political power

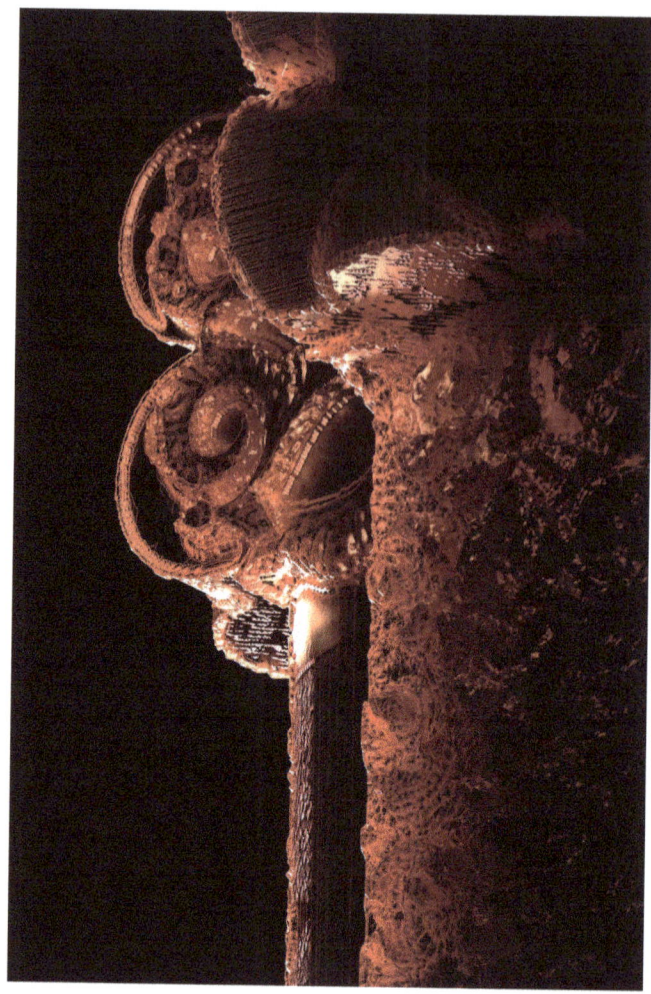

9 Choisy (wormseye) view of artifact

10 Close up view of artifact

struggles" come to light in a transmutable container that can now be inhabited, immersed by both water and ornament as they become one (Young 2021, 32). The process of this methodology makes material, decoration, and space one inexhaustable artifact of time. Lost in the vastness of water and consumed by its partness in space, ornament becomes renewed in the digital and cultural discourse of architecture.

"I sense Light as the giver of all presences, and material as spent Light. What is made by Light casts a shadow, and the shadow belongs to Light." — Louis I. Kahn [1]

ACKNOWLEDGMENTS

My deep appreciation to my thesis advisor Andrew Saunders, my dear friends Madison Green, Matthew Kohman, and Julianna Cano. My consultants Michael Young, Kivi Sotamaa, and Jawad Altabtabai. Brian De Luna and dear mentor Patrick Danahy who helped me develop the saplings that grew into the technical ambitions and curiosity of my thesis methodology. Patricia Guardiola and Kirby Bell from the Fisher Fine Arts Library for allowing me to access the fragments of the Alhambra under extraordinary circumstances. Marcos Sanchez, my instagram pal from Granada who generously photographed the cisterns during the Covid-19 pandemic. Alireza Borhani and Negar Kalantar, my first instructors in architecture school. Finally, my most special gratitude to my dear mentor, professor, and friend Gabriel Esquivel for helping me open doors to places I didn't know existed.

NOTES

1. Kahn often repeated statements such as this, and they can be found in various sources. See Louis I. Kahn, "Architecture, Silence and Light," Guggenheim Museum Lecture, New York, December 3, 1968. In *Louis Kahn; Essential Texts*, edited by R. Twombly. New York: W.W. Norton & Company (2003), pages 228–252.

11 Elevation view of artifact

12 Catalog of artifacts

REFERENCES

Bloomer, Kent 2000. *The Nature As Ornament*. New Haven: Yale University Press.

Jones, Owen 1856. *The Grammar of Ornament*. London: Day and Son.

Kubler, George 1962. The Shape of Time. New Haven: Yale University Press.

Nasr, Seyyed Hossein 1987. *Islamic Art & Spirituality*. New York: State University of New York Press

Young, Michael 2021. *Reality Modeled After Images*. London: Routledge.

IMAGE CREDITS

Figure 2: Owen Jones & Jules Goury
Figure 4.1: Marcos Sanchez
All other drawings and images by the author.

Paul Germaine McCoy is a Project Designer at Michael Hsu Office of Architecture and a Lecturer at the University of Texas at Austin School of Architecture. He holds a Bachelor of Environmental Design from Texas A&M University and a Master of Architecture from the University of Pennsylvania Weitzman School of Design. His portfolio, Protagonists of Architecture, received the 2020 Kanter Tritsch Prize in Energy & Architectural Innovation in concert with the 2020 Kanter Tritsch Medalist, Peter Eisenman. Upon graduating fron University of Pennsylvania, Paul received the Arthur Spayd Brook Memorial Prize: Gold Medal, following in the footsteps of former recipients Julian Abele, Louis I. Kahn, and Jenny Sabin. *Metropolis Magazine* has named Paul one of the Future 100 designers in the United States.

TESSERACT

Integrated Reconfigurable Autonomous Architecture System

Wanzhu Jiang, Jiaqi Wang, Tyson Hosmer, Ziming He

TESSERACT is an autonomous architecture developed through a voxel-based robotic material system that continuously reshapes communities through a socio-economic model with shifting fractional ownership. This incentivizes users to trade and share portions of physical space in real-time (Figure 1). Based on the Integrated Reconfigurable Autonomous Architecture System, TESSERACT buildings have a continuously adaptive lifecycle enabling the shifting spatial needs of communities to be negotiated through an Observe, Generate, [re]Assemble feedback loop (Figure 2). TESSERACT is implemented with three integrated components: an interactive platform, a space planning algorithm, and a distributed robotic material system.

The interactive platform is the information collection port observing both the shifting demands of inhabitants and environmental constraints (Figure 3). The environment interface processes dynamic environmental conditions through bitmaps and translates them into 3D data matrices. The user interface collects the community of users' requirements and characteristics such as desired space typologies and willingness to share and structures them as inputs driving the behaviors of the adaptive space generation algorithm.

The space planning algorithm is a multi-agent system that extends the principles of Stigmergic Space Adjacency Software (Meyboom and Reeves 2013), where each programmable agent represents an independent space

and communicates through its 3D voxelized environment. Spatial agents are trained with reinforcement learning to learn adaptive policies for adjusting their scales, shapes, and organization in relation to each other in response to changes in the environment and user requirements (Figure 4). Users' requirements are mapped into three collections of agent properties called "schema" that influence changes in their behavior. Relational Schema represented as "User Hue," defines the degree of relationality to neighboring agents of a similar or different color. Space Schema parameters (V, P,

PRODUCTION NOTES

Architect:	Jiaqi Wang, Wanzhu Jiang, Ying Lin, Zongliang Yu
Client:	Master's Thesis Project
Date:	2021

1   The perspective and section of TESSERACT.

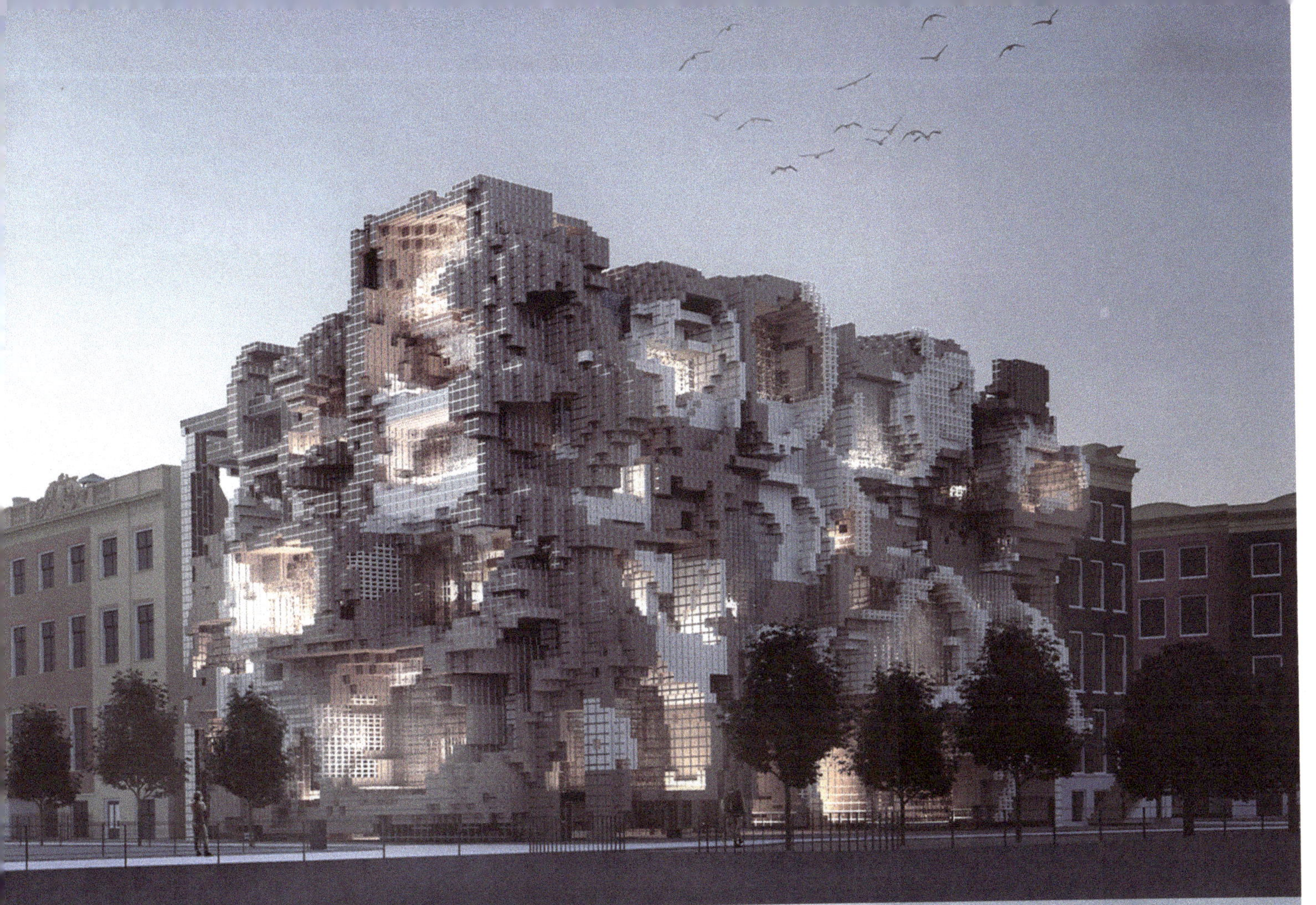

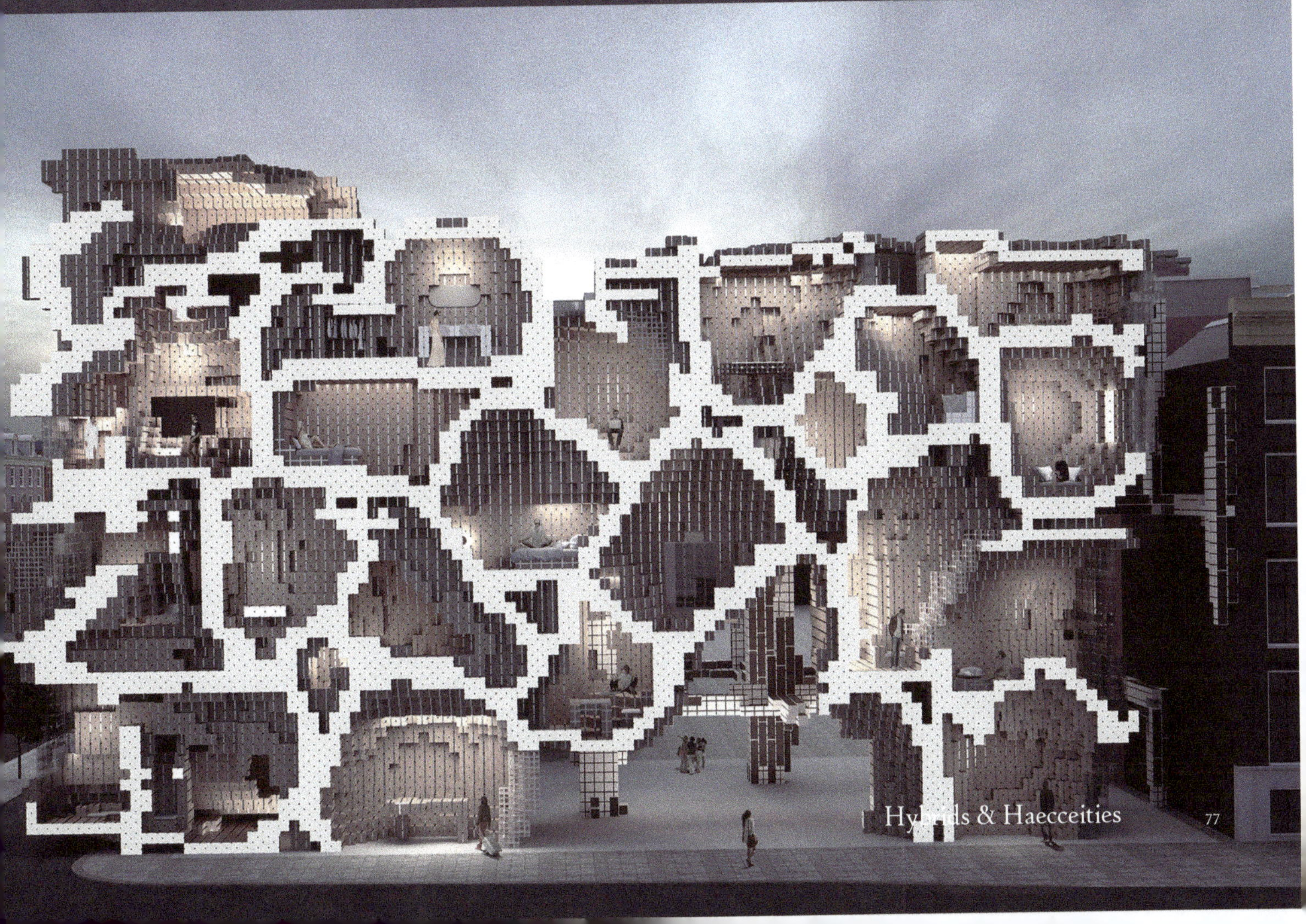

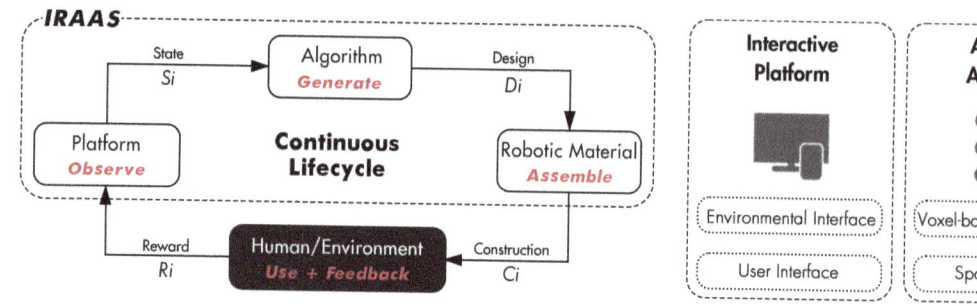

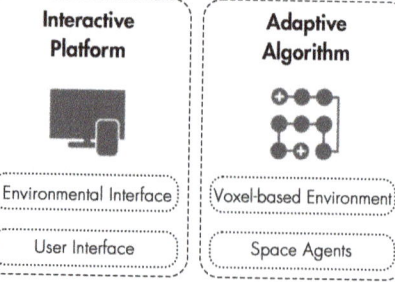

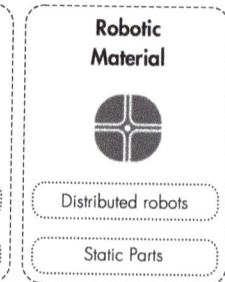

2   Continuous lifecycle and system composition

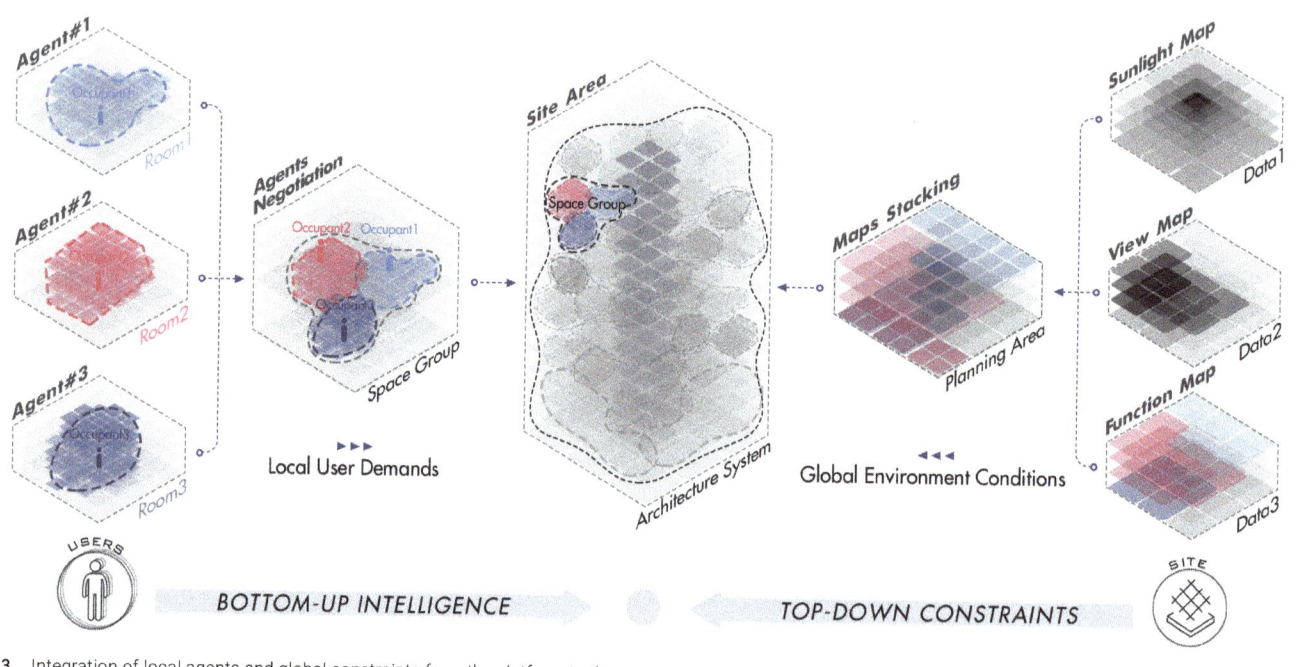

3   Integration of local agents and global constraints from the platform to the space planning algorithm

F) define the target space's volume, proportion, and form. Negotiation Schema parameters (I, C, E) define the agent's relationship with its neighbors, neighbor clusters, and the planning environment (Figure 4). Deep Reinforcement Learning with Self-Play was used to train the agents to maintain a mapping between their Space Schema goals and their adjusted volumes while negotiating their space in relation to each other through their Relational Schemas and Negotiation Schema parameters (Figure 4). Different types of agents were trained and tested through a series of experiments with 2 users and 5 agents (Figure 5). The trained agents were applied to the design of a station to get real-time results with different occupancies (Figure 6).

The distributed robotic material system was developed with a structured environment, where distributed robots slide their bodies on tracks built into passive blocks that enable their locomotion while utilizing a locking system of knobs to reconfigure the assemblages they move across (Figure 7).

The distributed robots have L-shaped bodies of three voxels with behaviors for sliding, changing direction, pushing and pulling, locking and unlocking static parts (Figure 7). Our custom robotic control system employs a wireless bi-directional communication protocol that connects the Unity 3D simulation environment to the physical environment, triggering the Dynamixel motors while returning motor sensor data. Our physical prototype testing demonstrated a series of locomotion and reconfiguration tasks (Figure 8). Self-play reinforcement learning was used to train multiple robots in the simulator to cooperatively reconfigure a wall from a default state through efficient sequences closely matching the series of goals (Figure 9).

TESSERACT is situated within a larger body of research undertaken at the Living Architecture Lab at the Bartlett related to developing autonomous architecture systems (Hosmer and Panagiotis 2019). By designing the three subsystems in relation to each other— with a cyber-physical

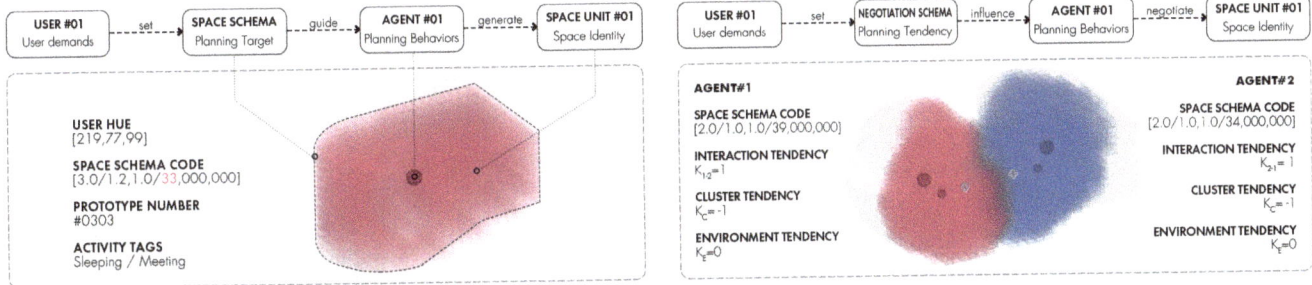

USER #01 — User demands — set — SPACE SCHEMA — Planning Target — guide — AGENT #01 — Planning Behaviors — generate — SPACE UNIT #01 — Space Identity

USER #01 — User demands — set — NEGOTIATION SCHEMA — Planning Tendency — influence — AGENT #01 — Planning Behaviors — negotiate — SPACE UNIT #01 — Space Identity

USER HUE
[219,77,99]

SPACE SCHEMA CODE
[3.0/1.2,1.0/33,000,000]

PROTOTYPE NUMBER
#0303

ACTIVITY TAGS
Sleeping / Meeting

AGENT #1

SPACE SCHEMA CODE
[2.0/1.0,1.0/39,000,000]

INTERACTION TENDENCY
$K_{12}=1$

CLUSTER TENDENCY
$K_C=-1$

ENVIRONMENT TENDENCY
$K_E=0$

AGENT #2

SPACE SCHEMA CODE
[2.0/1.0,1.0/34,000,000]

INTERACTION TENDENCY
$K_{21}=1$

CLUSTER TENDENCY
$K_C=-1$

ENVIRONMENT TENDENCY
$K_E=0$

4 Space schema and negotiation schema control the agent's occupation and negotiation behaviors

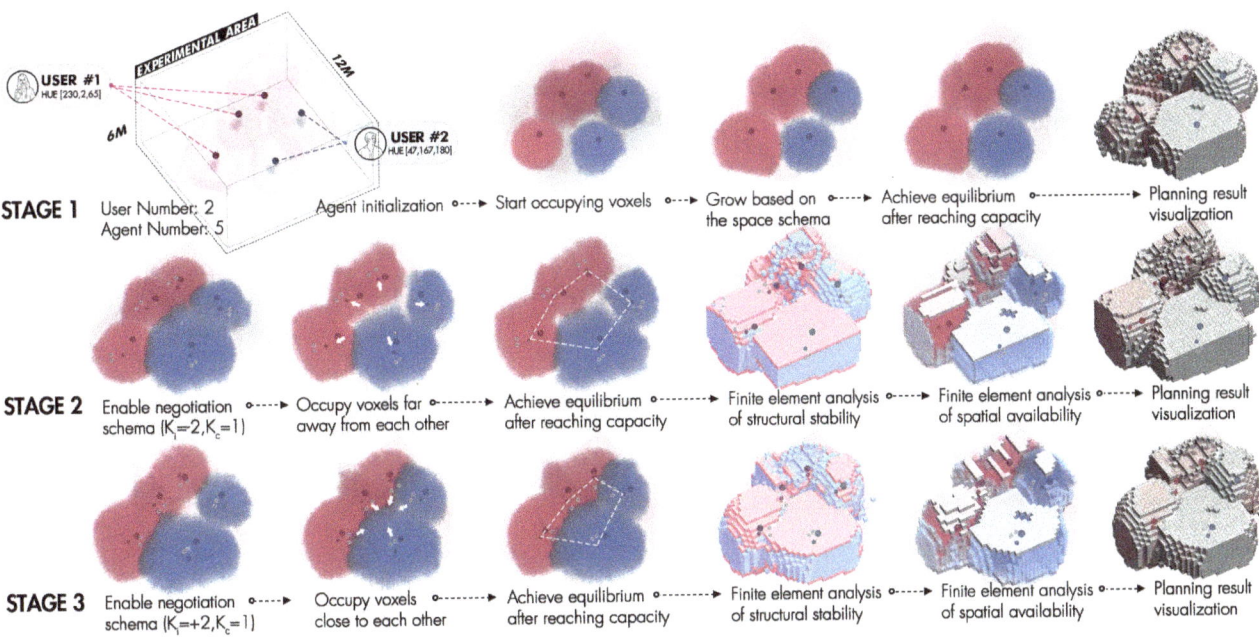

STAGE 1 — User Number: 2 / Agent Number: 5 — Agent initialization → Start occupying voxels → Grow based on the space schema → Achieve equilibrium after reaching capacity → Planning result visualization

STAGE 2 — Enable negotiation schema (K_I=2,K_c=1) → Occupy voxels far away from each other → Achieve equilibrium after reaching capacity → Finite element analysis of structural stability → Finite element analysis of spatial availability → Planning result visualization

STAGE 3 — Enable negotiation schema (K_I=+2,K_c=1) → Occupy voxels close to each other → Achieve equilibrium after reaching capacity → Finite element analysis of structural stability → Finite element analysis of spatial availability → Planning result visualization

5 Multi-user space negotiation experiments

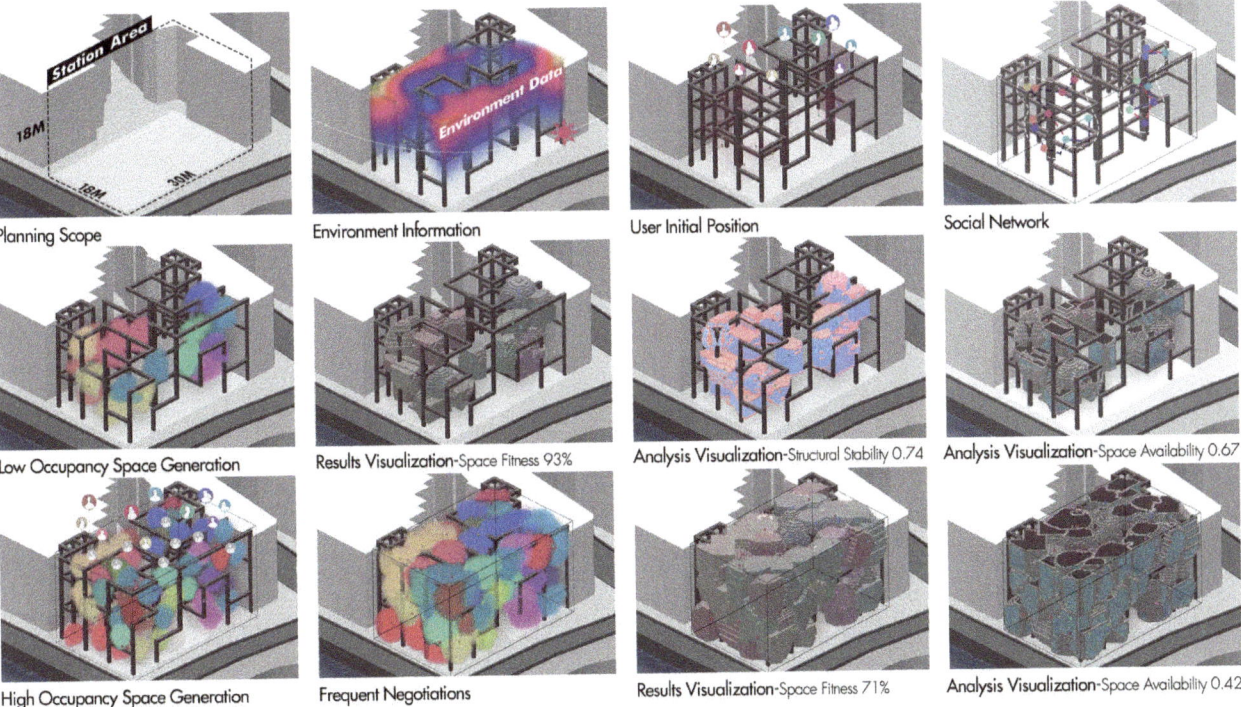

Planning Scope

Environment Information

User Initial Position

Social Network

Low Occupancy Space Generation

Results Visualization-Space Fitness 93%

Analysis Visualization-Structural Stability 0.74

Analysis Visualization-Space Availability 0.67

High Occupancy Space Generation

Frequent Negotiations

Results Visualization-Space Fitness 71%

Analysis Visualization-Space Availability 0.42

6 The generation process of TESSERACT's example station in different occupancy rate

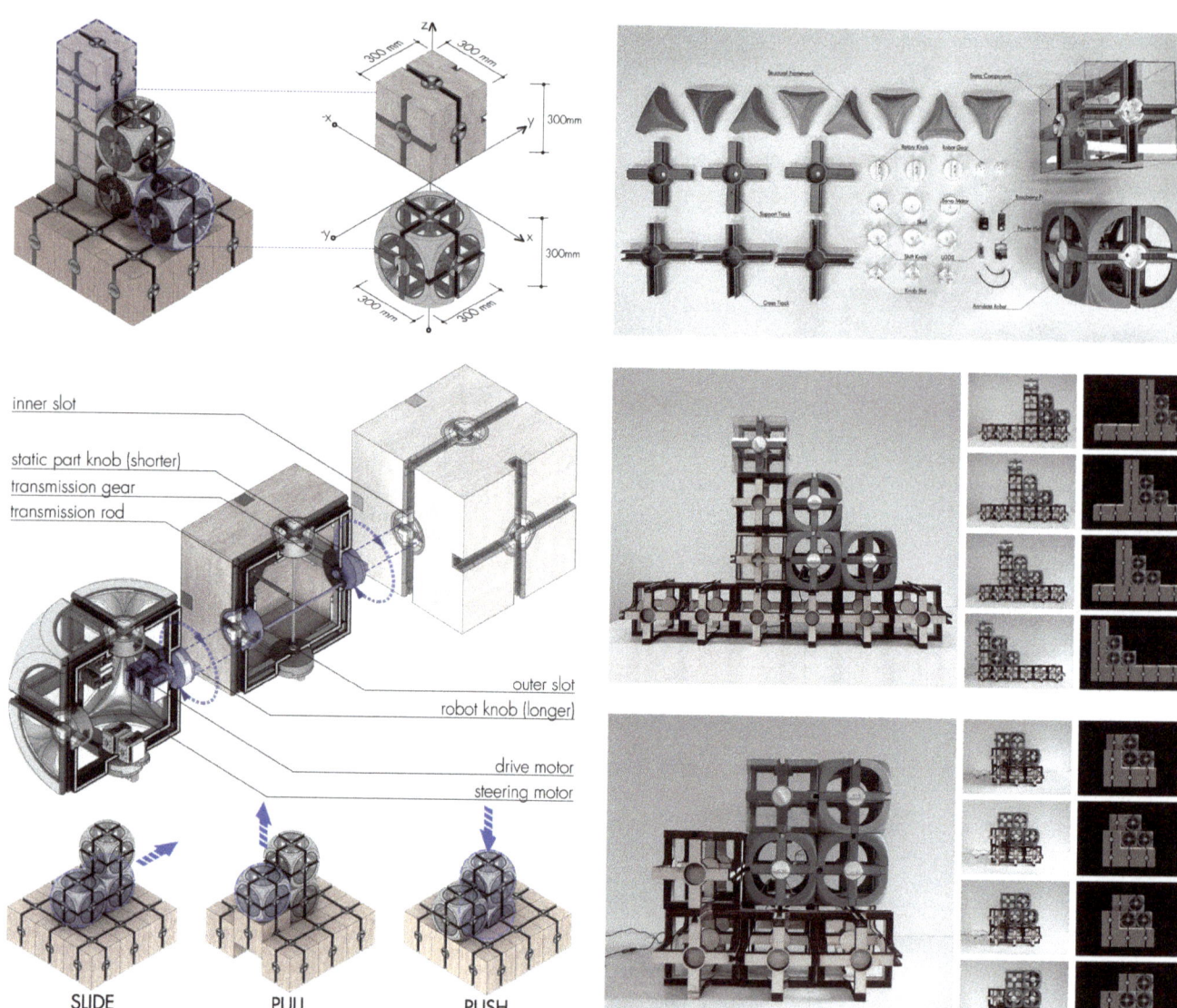

7 Static parts and distributed robots, joint details, and action examples

8 The physical prototype tests: locomotion and reconfiguration

control protocol that considers both the constraints of the AI-driven space planning algorithm and the robotic material system—this research has potential for autonomous physical adaptation with a continuous feedback loop. The project further speculates on the potential for this to fundamentally change our relationship to our built environment through its socio-economic proposition.

ACKNOWLEDGMENTS

This Project is conducted in UCL, The Bartlett AD Research Cluster 3 (RC3), Living Architecture Lab. Team members: Jiaqi Wang, Wanzhu Jiang, Ying Lin, and Zongliang Yu. Tutor: Tyson Hosmer, Octavian Gheorghiu, Philipp Siedler, Panagiotis Tigas, Ziming He, and Barış Erdinçer.

REFERENCES

Hanna, Sean. 2020. "Architecture as Agent." Online publication. Accessed September 8, 2022. https://www.researchgate.net/publication/341160906_Architecture_as_agent.

Hosmer, Tyson, and Panagiotis Tigas. 2019. "Deep Reinforcement Learning for Autonomous Robotic Tensegrity (ART)." In *ACADIA 2019: Ubiquity and Autonomy; Proceedings of the 39th Annual Conference of the Association for Computer Aided Design in Architecture (ACADIA)*, edited by K. Bieg, D. Briscoe, and C. Odom. Austin, TX: ACADIA.

Meyboom, AnnaLisa, and Dave Reeves. 2013. "Stigmergic Space." In *ACADIA 2013: Adaptive Architecture; Proceedings of the 33th Annual Conference of the Association for Computer Aided Design in Architecture*, edited by P. Beesley, O. Khan, and M. Stacey.

Tibbits, Skylar. 2012. "Design to Self-Assembly." *Architectural Design* 82 (2): 68–73.

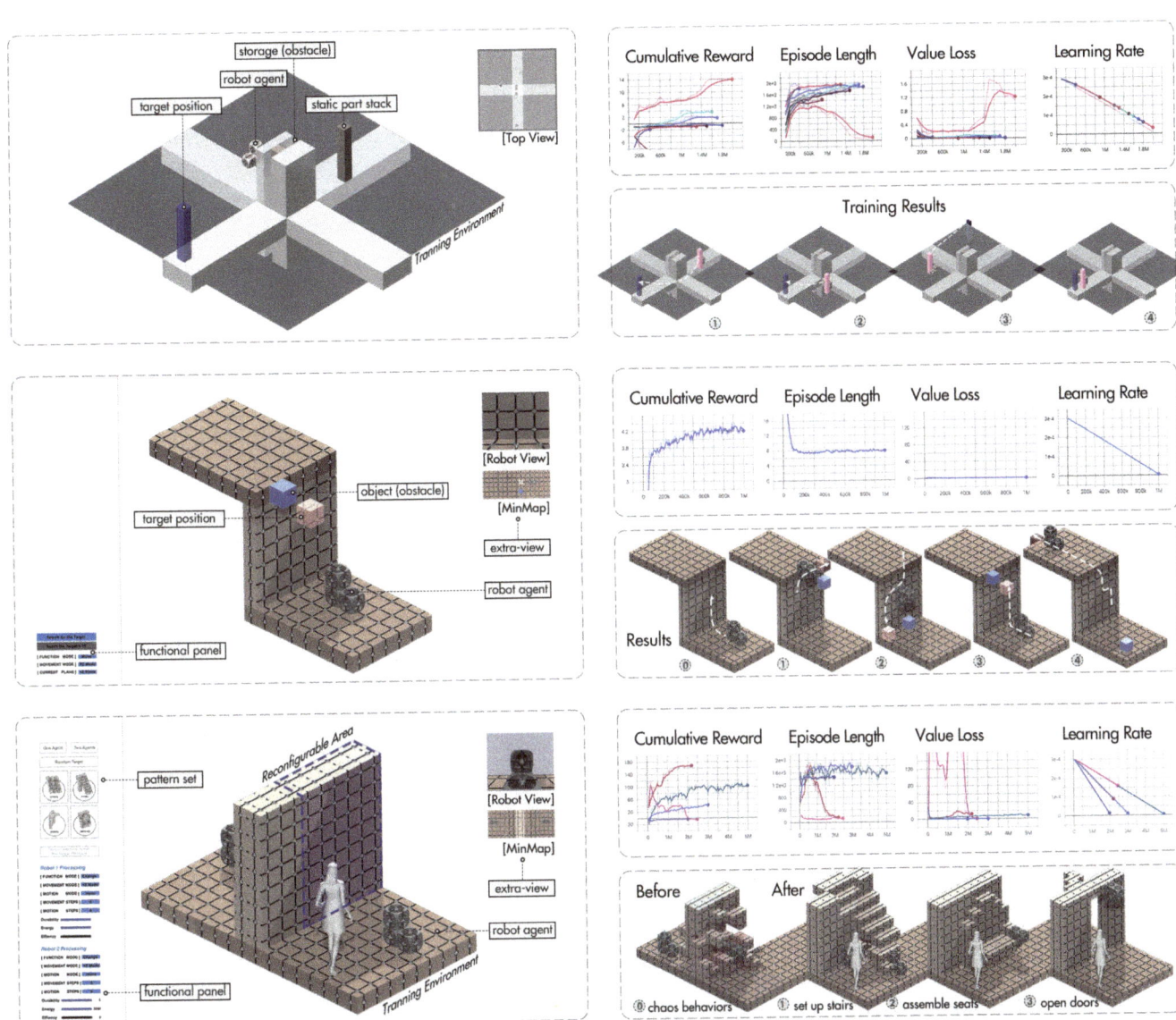

9 Three stages of the deep reinforcement learning: (1) 2D transporting, (2) 3D path finding, and (3) cooperatively design goal assembling

IMAGE CREDITS
All drawings and images by the authors

Wanzhu Jiang is a PhD student and researcher at the South China University of Technology, whose research direction is artificial intelligence and autonomous architecture. She received her master's degree from UCL in Bartlett BPro RC3 in 2021, supervised by Tyson Hosmer.

Jiaqi Wang is a PhD student and researcher at the South China University of Technology, focusing on artificial intelligence aided architectural design, generative design, and robotics. He received his master's degree from UCL in Bartlett BPro RC3, guided by Tyson Hosmer.

Tyson Hosmer is an architect, researcher, and software developer working at the intersection of design, computation, AI, and robotics. He is the Director of the Architectural Design masters program and a Lecturer at the Bartlett School of Architecture in London, where he directs the Living Architecture Lab (RC3). He is a Senior Associate Researcher with Zaha Hadid Architects, leading grant-funded research development of cognitive agent-based technologies and machine learning for design. His 15 years of experience in practice include working in Asymptote Architecture, Kokkugia, AXI:OME, and serving as Research Director with Cecil Balmond Studio for over six years. Tyson was previously a Course Tutor with the AADRL for seven years and has been a visiting professor in several institutions internationally.

Ziming He is a designer and a software developer focusing on computational design, generative design, architectural visualization, and robotics. He is currently working as a lead designer in Zaha Hadid Architects and is a technical tutor with RC3, Bartlett School of Architecture, UCL

The Serlio Code

Beyond Classic Language

Gabriel Esquivel, Jean Jaminet, Shane Bugni

The discourses on language and drawing established by the Classical architectural treatise find new disciplinary relevance in current advancements and discussions concerning machine learning. This research project examined the illustrated expositions of Sebastiano Serlio through the lens of artificial intelligence. The intention of this project was to use Serlio's illustrations to modulate their qualities and problematize their 2D to 3D translation beyond the rules of representation and orthographic projection. Three operative models were presented: columns, plans and facades, and porticos—developed by augmenting and interpreting layered generative adversarial networks that drive an integrated parametric 3D process. These insights and investigations disclosed alternative theoretical connections between information processing and aesthetic communication as well as emerging modes of creative digital production.

For the purposes of this project, we negotiated between parametric and manual integration in the modeling process. The relation with Serlio's work was understood through to the power of data, and Serlio's coded operations of orthographic projection drawing disclose other aesthetic and formal logics for architecture. However, at other times architectural elements are misaligned, suggesting that a latent diagrammatic operation is at work, which is a new code.

PRODUCTION NOTES

Architects: Gabriel Esquivel, Jean Jaminet, Shane Bugni

Status: Built

Location: College Station, Texas

Date: 2021

1 *The Serlio Code*, installation view, Wedge Gallery, Woodbury University (Zelig Fok 2022)

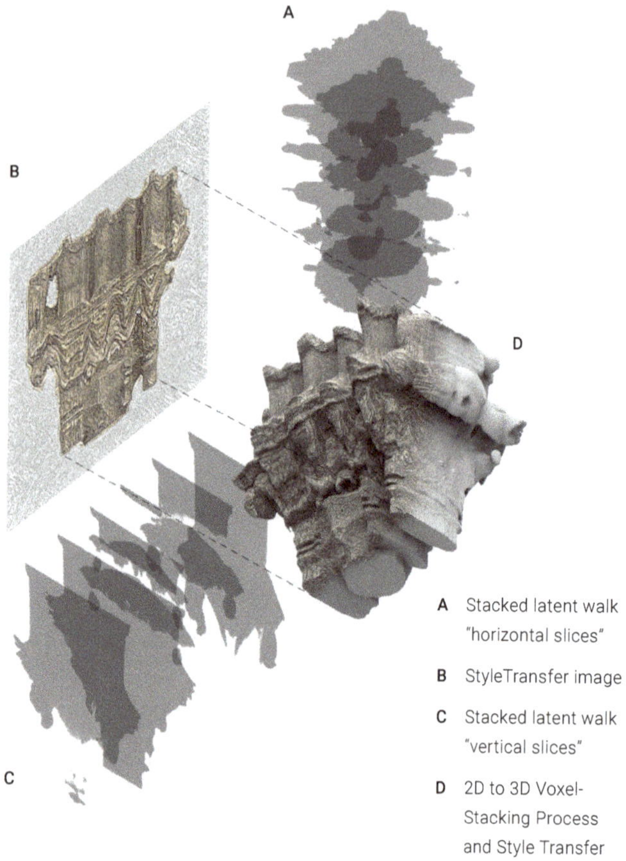

A   Stacked latent walk
    "horizontal slices"

B   StyleTransfer image

C   Stacked latent walk
    "vertical slices"

D   2D to 3D Voxel-
    Stacking Process
    and Style Transfer

2   2D-3D Process and StyleTransfer

3   Final object

4   Rendered portico

5   Preliminary study before use of SinGan

6   Preliminary study before use of SinGan

7   Displacement map

COLUMNS AND FACADES

Using the logic of orthographic projection inherent to the column as a drawing and its tectonic method of assembly, the 3D model was re-encoded by taking the horizontal and vertical cross-section. These new encoded patterns became the data set to be trained by an image-based neural network. Once these patterns were realized through the use of styleGAN models, new 3D models relying on the voxel resolution to determine the thickness of each layer were generated from the frames of the animated latent walk.

The articulation of the model was further explored by UV mapping and style transfer to move beyond the limitation of perspective and parallel projections (Figure 3). The style transfer procedure was deployed using the unrolled UV curvature map of the morphed column capital object as the content image, while a Serlian detail image supplied the style information. The object was then re-wrapped with the new transferred texture and its paired displacement map. The displacement is baked onto the geometry, supplying the object with the new style transfer information (Figure 4).

PORTICOS

Each image of a Serlian portico was broken down into its recognizable elements, such as the pediment, then reinterpreted and fragmented by AI (Figure 5). Unlike the previous columns and plan-façade experiments, a portion of the generalized portico was additionally trained using sinGAN, which fragments the elements to generate a new set of profile frames (Figure 5). The processed image reveals alternative logics of tectonics and assembly. These AI-generated image fragments were processed through image manipulation and 3D sculpting software to produce geometry from the black and white pixel values. These new relationships of part-to-part expand the idea of a portico as each recognizable element becomes disjointed and estranged from its original qualities through the processes of AI fragmentation and reassembly via the morphing of the voxelized fragments (Figures 4 and 5).

FABRICATION

We used different types of foam as our base material, and since a 3-axis mill does not have the ability to make

8 Fabricated portico and configuration

undercuts, the biggest challenge was to develop a technique that allowed the piece to be articulated on both the front and back of the construct. A parameterized sectioning technique was used to split the piece into 4-inch vertical sections. Using this technique, it was possible to mill many small pieces of the overall form that fit together. Undercuts were not completely unavoidable, but we tried to avoid them as much as possible via sectioning, and the parts where undercuts could not be avoided had to be sanded by hand to approximate the geometry (Figures 6, 7, and 8).

ACKNOWLEDGMENTS

We would like to thank the Texas A&M Advanced Research team for their partipation: John Scott, Brendan Bjerke, Austin White, Spencer Young, Ana Rico, Nathan Gonzalez, Erin Carter, and Luis Sanabria.

IMAGE CREDITS

Figure 1: Zelig Fok, 2021
Figure 2, 3, 4: John Scott, Luis Sanabria, 2021
Figure 5: Nathan Gonzalez, 2021
Figure 6: Spencer Young, 2021
Figure 7: Nathan Gonzalez, 2021
Figure 8: Austin White, 2021
Figure 9: Luis Saucedo, 2021

Gabriel Esquivel is Associate Professor at Texas A&M Univeristy and the director of the T4T lab and AI Advanced Research Lab at Texas A&M University. Gabriel is a native of Mexico City, where he was educated as an architect at the National University. He received his Master's degree in Architecture from Ohio State University. He previously taught Architecture and Design at the Knowlton School of Architecture and the Design Department at The Ohio State University. He is founding partner of the online magazine *Agencia*, a publication dedicated to problems of teaching theory and technology in Mexico.

Jean Jaminet is Assistant Professor at Kent State University College of Architecture and Environmental Design. He holds a Master of Architecture from Princeton University and a Bachelor of Science in Architecture from The Ohio State University. His prior teaching appointments include Visiting Assistant Professor at the University of Arkansas, Instructor at the Savannah College of Art and Design, and Five College Visiting Lecturer in Architectural Studies.

Shane Bugni is a graduate student at The Weitzman School of Design. He is a former student at Texas A&M University, where he co-taught the Advanced Research Lab on Artificial Intelligence and Robotics from 2020-2022 with Gabriel Esquivel. He has taught several workshops like Digital Futures 2021 and 2022.

9 Fabricated Portico Configuration.

Voxel Cloud

Volumetric Scaffolding in 3D Pixel Space

Julian Edelmann

The project *Voxel Cloud* is an experimental prototype that deals with deep and vague tectonics in an architectural context by investigating resolution and complexity.

The project confronts the exuberance of complex geometries generated by algorithms with the perception of humans, thus it questions the central role of the human within this process by attempting to blend nature and technology. The end result in form of a building application (Figure 5) is not a proposal per se, but rather a speculation how data and computation can generate an architecture that can be build by machine and inhabited not just by humans, but also by micro to macro organisms in a post-anthropocentric environment.

HIGH-RESOLUTION ARCHITECTURE

This speculation positions itself within the discussion of high-resolution architecture and concerns a voxel-based lattice technique as data-materialization and direct translation between digital and physical via robotic additive manufacturing.

PRODUCTION NOTEs

Designer:	Julian Edelmann
Type:	Bachelor's Degree Thesis
Institution:	i.sd - Structure and Design
Affiliation:	University of Innsbruck
Date:	2021

1   3d printed physical artefact

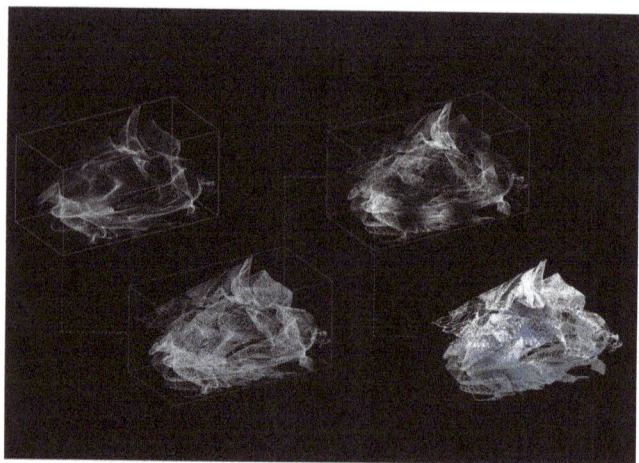

2   Digital process of pointcloud solidification

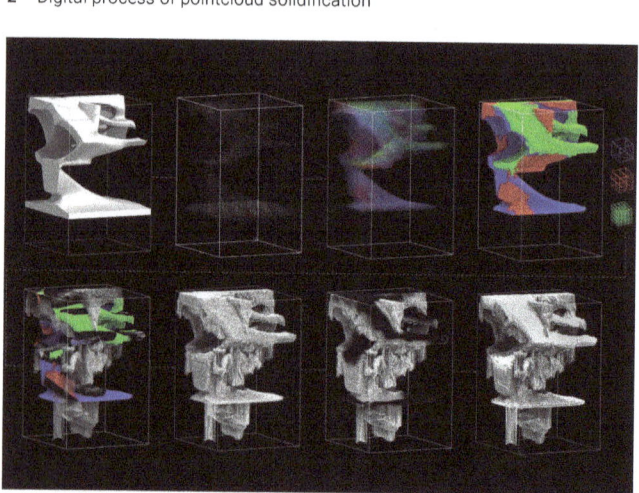

3   Volumetric modeling technique

4   Bidirectional Evolutionary Structural Optimization (BESO)

Therefore, it aligns alongside the lineage of research projects by Alisa Andrasek, Biothing and Wonderlab research, UCL, e.g "Cloud Pergola, Venice Biennale 2018" (Andrasek 2018), where a voxel grid was informed and controlled by noise in order to generate porous and poly-scalar structures which were 3D printed by utilizing realtime adaptive print path generation trained by AI.

Another prototype to mention here is the "Voxel Chair 1.0, Centre Pompidou 2017," by Manuel Jimenez Garcia and Gilles Retsin, developed within the research of the Design Computation Lab, UCL. It follows a similar volumetric modelling approach, by representing a chair as a 3-dimensional pixel space that is translated into a combination of line fragments for robotic extrusion.

DIGITAL PROCESS

The project is utilising a computational workflow, whereby scanned pointcloud data act as inputs for the design tool.

This three-dimensional dataset serves as a digital datascape that can be further informed by local and global attributes, such as environmental conditions, structural loads, digital fabrication parameters, and material constraints (Figure 2).

The pointcloud was transformed into a uniform voxel grid, where a Bidirectional Evolutionary Structural Optimization algorithmn (Young, Querin, Steven, and Xie 1999, 188) analyzed the structure on internal forces (Figure 4). By that process, the structure was separated into areas with different voxel densities in order to make it more efficient (Figure 3). Those differentiated areas were futher articulated by noise for the sake of aesthetics.

This process further transforms the pointcloud into a voxel cloud, whereas each voxel acts as a container of unique information and data at its specific location. This datascape, which is mostly hidden and invisible to humans, can

5   Building application study.

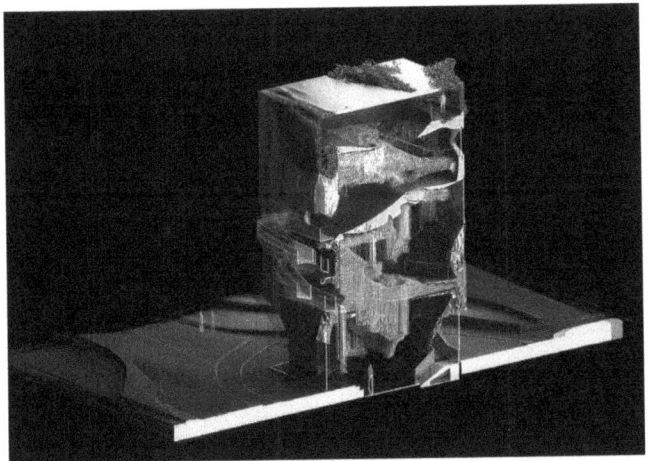

6   3d section building application study.

7   Voxel scaffold with several densities and articulations.

8   Circulation generated by voxels.

then be decoded and fabricated by the machine (Jimenez Garcia, Soler, and Retsin 2017, 149).

As output, the design tool apparatus produces a filigree, lightweight, data-informed structure (Figure 9) that has a high variety in density and articulation. It is characterised by transitions between solid and fibrous as well as order and disorder.

Physical artefacts of this structure (Figure 10) were made to test the materialisation of this volumetric modeling workflow by using Binder Jet 3D print technologies (Hansmeyer, Dillenburger, and Mathias 2018, 394).

BUILDING APPLICATION

The digital workflow was then applied to a concrete building application (Figure 5). There is no clear differentiation between interior or exterior, but rather several densities that can enhance different microclimate conditions (Figure 7).

The voxel scaffold acts as a canvas for plants and animals to adapt and build on as well as to guide water throughout the whole structure. It is located in a park in Innsbruck, Austria, where it can get overgrown by time and blend with the natural surroundings.

ACKNOWLEDGMENTS

This project was part of the author's Bachelor thesis completed at i.sd institute for structure and design at the University of Innsbruck. The author would like to acknowledge to Prof. Kristina Schinegger, Prof. Stefan Rutzinger, Viktoria Sandor, and Klemens Sitzmann for supervising this project, as well as Leiber Group GmbH & Co.KG for 3D printing support.

9   Detail of the voxel based lattice scaffold

REFERENCES

Andrasek, Alisa. "Cloud Pergola." Accessed July 01, 2022. https://www.alisaandrasek.com/projects/cloud-pergola.

Jimenez Garcia, M., V. Soler, and G. Retsin. 2017. "Robotic Spatial Printing." In *ShoCK! - Sharing Computational Knowledge!; Proceedings of the 35th eCAADe Conference*, Volume 2, edited by A. Fioravanti, S. Cursi, S. Elahmar, S. Gargaro, G. Loffreda, G. Novembri, and A. Trento. Sapienza University of Rome, Rome, Italy. 143–150. https://doi.org/10.52842/conf.ecaade.2017.2.143.

Young, V., O.M. Querin, G.P. Steven, et al. 1999. "3D and multiple load case bi-directional evolutionary structural optimization (BESO)." *Structural Optimization* 18, 183–192. https://doi.org/10.1007/BF01195993

Bernhard, M., M. Hansmeyer, and B. Dillenburger. 2018. "Volumetric modelling for 3D printed architecture." In *Advances in Architectural Geometry 2018*, edited L. Hesselgren, A. Kilian, S. Malek, K. Olsson, O. Sorkine-Hornung, and C. Williams. Vienna 2018. 392–415.

IMAGE CREDITS

All images by the author.

Julian Edelmann is currently pursuing a Master in architecture at the University of Innsbruck. He works as a research assistant at REX|LAB robotic experimentation laboratory at the Department of Experimental Architecture, studio Marjan Colletti. Prior to that he worked as a Lecturer at i.sd, institute for structure and design at the University of Innsbruck.

10 Silicia sand model using Binder Jet 3D print technology

Hybrids & Haecceities   93

Additive Hyper-Ornamental Prototypes

Surface Articulation as Structural Leverage in 3D Printing

Pavlos Fereos, Eftychios-Nicolaos Efthimiou, Kilian Bauer, Julian Edelmann

This project presents two experimental prototypes built using a 6-axis Cobot (Universal Robots UR10e collaborative robot) and PETG (polyethylene terephthalate glycol) filament processed by a plastic extruder (Herz Robot 0.8).

The aim was to incorporate intricate design elements into 3D models to test or even increase the material's structural abilities and to 3D print large and highly articulated architectural mock-up models on a 1:1 scale.

Two design proposals were put forth, each posing different challenges, as described below.

COLUMN PROTOTYPE

For the first prototype the intention was to print an ornamental 3 meter tall column (Figure 4), which does not carry any load except its own weight; to challenge the material, the design is slightly tilted to increase tension and static loads.

Inspired by corinthian columns, the design is structured into a wide base, followed by a filigree shaft with a highly ornamental capital. A gradient effect aims to unify the ornamental capital with the shaft and base. For fabrication constraints (printing time and robot bounding box), the column was separately printed in five different parts with a height of 60 cm (Figure 2), making the accuracy of each print important in order to secure their connection afterwards.

PRODUCTION NOTES

Laboratory:	Rex|Lab
University:	University of Innsbruck
Status:	Build
Location:	Innsbruck
Date:	2021 - 2022

1   Final 3d printed throne

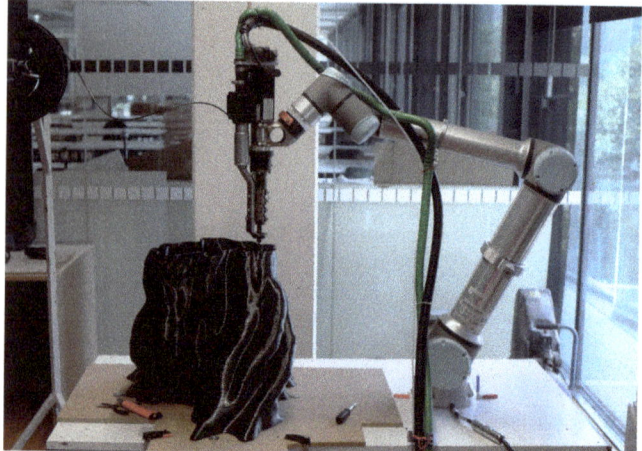

2 Planar printing of a column segment

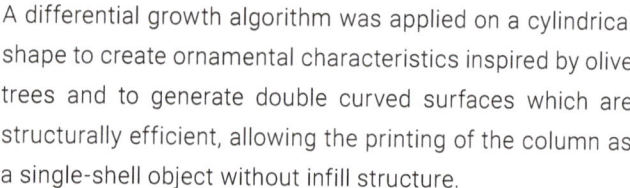

3 Transition between base and column

4 Final assembled 3m high column

A differential growth algorithm was applied on a cylindrical shape to create ornamental characteristics inspired by olive trees and to generate double curved surfaces which are structurally efficient, allowing the printing of the column as a single-shell object without infill structure.

The geometry was sliced into horizontal 1mm print path layers to allow maximum resolution, but also to better deal with cantilever areas. To fix the 3D printed parts onto a wooden sheet, the first layer includes lugs to fasten the prints and prevent warping. The five prints were assembled and glued together, whereby the interior lugs of the bottom-most part were printed eight layers high and were used to screw the column onto a wooden base which counterbalances the weight of the inclined design (Figure 3).

THRONE PROTOTYPE

As a next step, the aim was to fabricate a not purely decorative but usable object that could carry loads (Figures 1, 8, 9).

A throne design was chosen as a case study because, as a seating object, it has to handle the weight of a person (Figure 9), but also a throne has to have ornamental qualities which can contribute to the rigidity of the 3D printing paths.

The design of the throne was structured into three different goals: Overall shape, surface articulation, and structural infill. The overall shape was polymodeled and informed by the 1.3 meter radius bounding box of the robot, aiming to be printed in one seamless piece.

The double-curved surface technique, tested in the column prototype, was applied, this time by modeled wrinkles (Figure 7), which enabled the efficient printing of the throne as a single-wall object. The surface articulation was elaborated

5 Infill structure underneath the seat

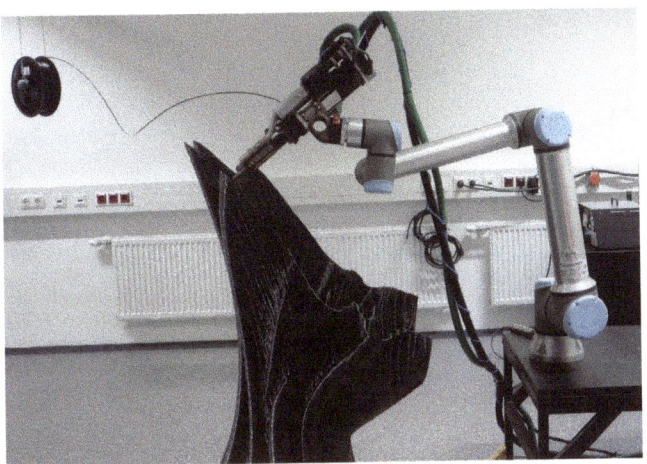

6 Non-planar printing process

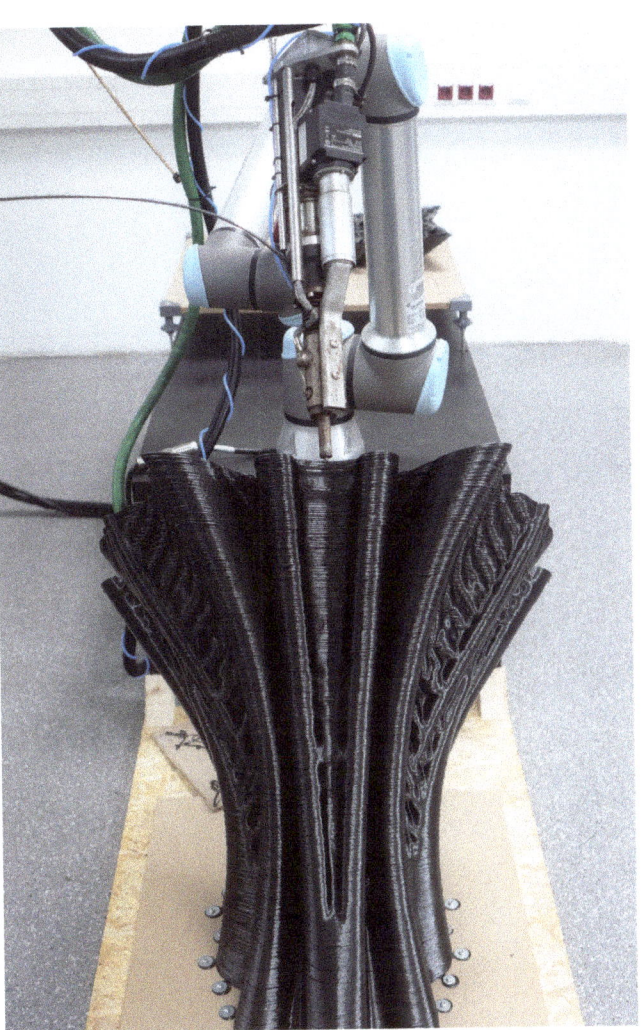

7 Deep wrinkles for rigidity

by a particle simulation that was informed by the wrinkle guide curves and the curvature of the overall shape, thereby provoking an ornament that enhances the geometrical attributes of the throne.

The structural infill (Figure 5) is a single curve path printed together with the outer shell with several welded points within the shell and between the infill and the outer shell to increase stiffness. The infill is embedded underneath the seating area, and a dense seating part for more comfort was printed separately (Figure 10).

The geometry is sliced non-planarly (Figure 6) with a variable layer height (0.7mm to 2mm) and fabricated with variable robot speed (larger layer height means slower robot speed).

CONCLUSION

While a vertical structure like a column works fine with horizontal print paths, this technique is very limited. However, the non-planar printing technique takes advantage of the 6-axis robot and therefore can be printed in 3D space (Jimenez Garcia and Retsin 2015, 335) to better deal with cantilevered areas.

ACKNOWLEDGMENTS

This research belongs to the robotic experimentation lab (REX|LAB) at the Department of Experimental Architecture at the University of Innsbruck, founded and directed by Prof. Dr. Marjan Colletti in October 2012 as a space for formal and material 1:1 scale experimentation and fabrication strategies. The project was completed within a digital fabrication seminar and is a collaboration between the University of Innsbruck and the University of Nicosia. The authors would like to acknowledge the following students for their contribution in this research: Marija Evdoridis, Sofoklis Kontakis, Lara Penz, Charlotte Thorn, and Daniel Ticker.

8 Throne seating area

REFERENCES

Jimenez Garcia, M. and G. Retsin. 2015. "Design Methods for Large Scale Printing." In *Real Time: Proceedings of the 33rd eCAADe Conference*, Volume 2, edited by B. Martens, G. Wurzer, T. Grasl, W.E. Lorenz, and R. Schaffranek. Vienna University of Technology, Vienna, Austria. 331–339. https://doi.org/10.52842/conf.ecaade.2015.2.331.

IMAGE CREDITS

All images by the authors.

Pavlos Fereos is an architect with a Diploma and an M.Sc from NTUA along with an M.Arch (DRL) from the AA. He is pursuing a PhD in Architecture. He holds the position of Senior Scientist at Exp.Arch Robotic Lab in Innsbruck. He is also a lecturer and program coordinator at the MSc in Computational Design and Digital Fabrication at University of Nicosia.

Eftychios-Nicolaos Efthimiou is a biologist and a designer, with a diploma from the University of Ioannina and a M.Arch GAD from the Volos School of Architecture, UTh with distinction. He is a Visiting Lecturer at Exp.Arch, Fakultät für Architektur, UIbk, Adjunct Faculty at the Department of Architecture, UNic, and a Scientific Committee Member at the Advanced Design Graduate Program, School of Architecture, AUTh.

Kilian Bauer is currently pursuing a Master in architecture at the University of Innsbruck. He works as a research assistant at REX|LAB robotic experimentation laboratory at the Department of Experimental Architecture, studio Marjan Colletti. Before that he acquired practical experience in construction as a trained architectural draftsman.

Julian Edelmann is currently pursuing a Master in architecture at the University of Innsbruck. He works as a research assistant at REX|LAB robotic experimentation laboratory at the Department of Experimental Architecture, studio Marjan Colletti. Prior to that he worked as a Lecturer at i.sd, institute for structure and design at the University of Innsbruck.

9 Person sitting on the throne

Versatile Bracketry

Contemporary Fabrication Techniques for Traditional Korean Architecture

Yong Ju Lee

GEO-LOCAL CONTEXT IN HISTORY

Versatile Bracketry is an architectural experiment employing algorithmic design technology and 3D printing, manipulating Gong-po—a wooden bracket element found in traditional Korean architecture. Although there has been some recognition and reflection toward the inclusion of traditional forms in modern design, the mainstream in Korean architecture has been Western-oriented. However, advanced computation technology provides both a new perspective and approach in this field, and higher productivity and efficiency.

This experiment suggests a modern construction methodology studying Gong-po, a symbolic element in structural and ornamental aspects. Gong-po distributes or concentrates the weight of the roof to serve a structure-buffering function. Not only does it expand the interior space and elevate the building for grand appearance, but also creates

PRODUCTION NOTES

Architect:	Yong Ju Lee
Status:	Permanent Exhibit
Location:	SeoulTech Arch Gallery, Seoul, South Korea
Date:	2021

1 Overall shape of *Versatile Bracketry* (Seungwoo Kim, 2021)

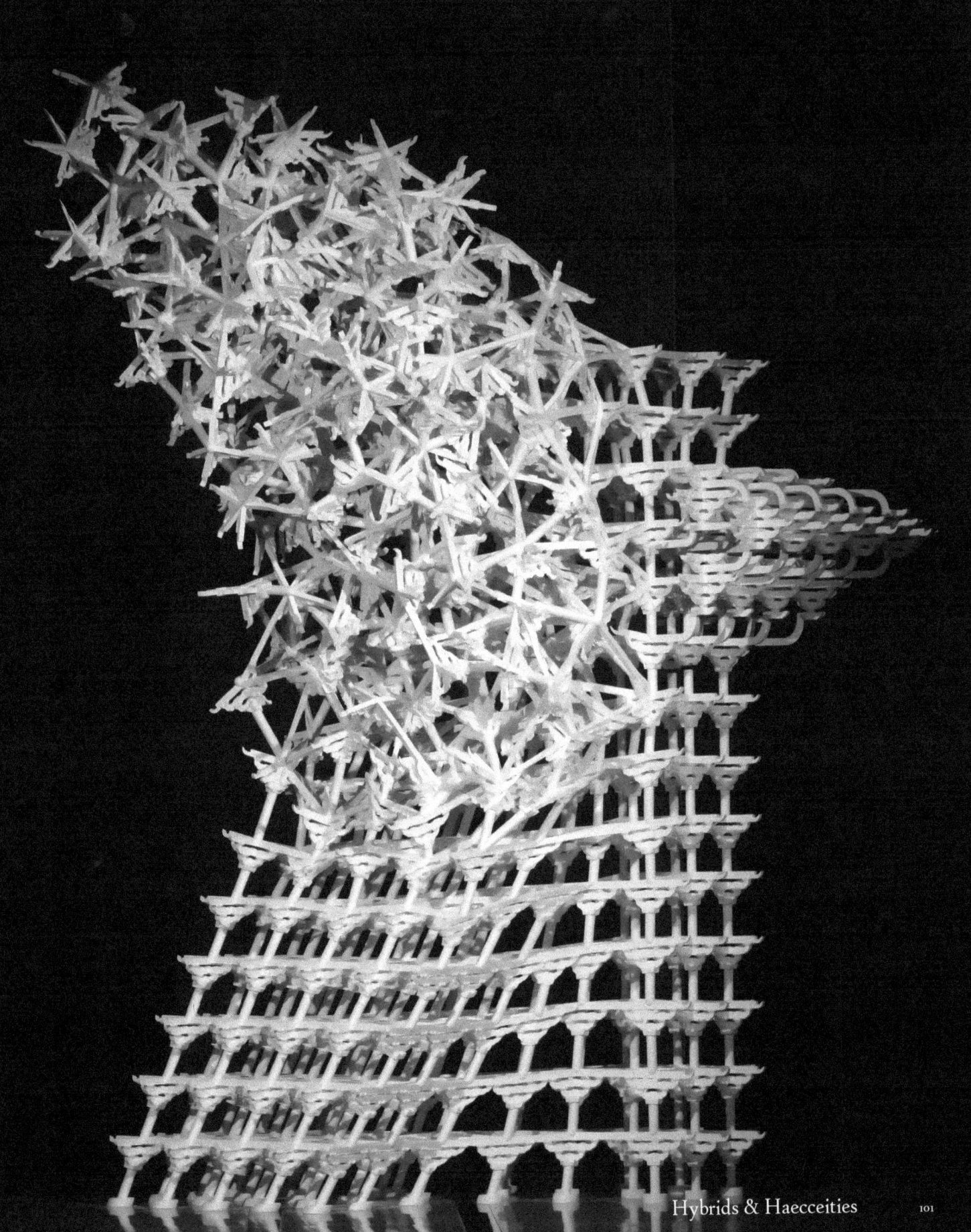

2 *Versatile Bracketry* in human scale (Seungwoo Kim, 2021)

a delicate and splendid the composition and artisanship are delicate and splendid, which eventually create an lending the structure an important aesthetic character. Therefore, In traditional architecture, Gong-po is the most important decorative expression of the a building, and it has its own uniqueness features based on each specific historical eras.

GEOMETRIC DISCIPLINE

This project employs Ju-sim-po, one type of Gong-po from Buseoksa, a Buddhist temple built in 1562. Constructed in wood, Buseoksa has never been disassembled since the initial installation and is well preserved; as a structure, this temple is considered sufficiently representative of traditional Korean architecture.

Keeping the constraints of relation and shape of the traditional structure, a vertical series of three-layered spreading brackets in Gong-po serve as the geometric basis of the project algorithm. Through Grasshopper, the graphic algorithm add-on for Rhinoceros, a system is set that utilizes a triangular shape of the nodes of the original composition. Handling the guide triangle produces an efficient strategy to generate various types of rigid Gong-po system. A systematic tool in the algorithm can give variability to the size of Gong-po, the angle of Chum-cha (diagonal supporter and connector in Gong-po), and the composition of Chum-cha, therefore transforming elements to suit the needs of the designer. Each component keeps its independent identity in the three-dimensional network, and– the connectivity between and among representations can be manipulated and expanded by the emerging persistence of network computation In this design methodology, Gong-po elements are accumulated in four different types in a continuous spectrum for a built human-scale structure.

BUILT PROTOTYPES

For the wooden prototype, interconnected joinery system parts are controlled by Boolean expression. By manipulating

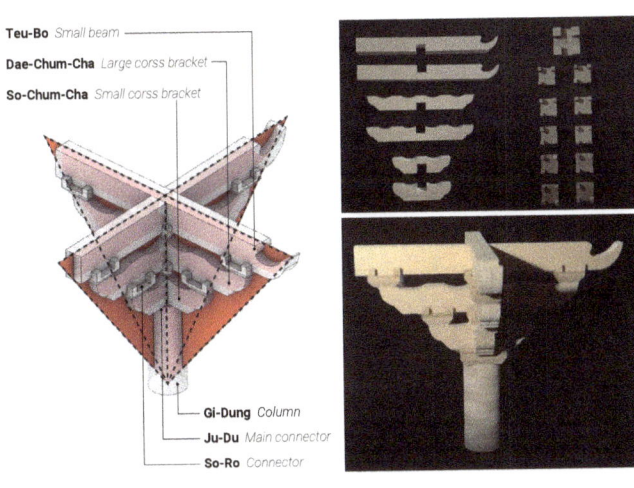

- **Teu-Bo** *Small beam*
- **Dae-Chum-Cha** *Large corss bracket*
- **So-Chum-Cha** *Small corss bracket*

- **Gi-Dung** *Column*
- **Ju-Du** *Main connector*
- **So-Ro** *Connector*

3   Gong-po: standandard type with CNC-fabricated prototype in wood

5   Gong-po in Buseoksa (©Province of Gyeongsangbuk-do)

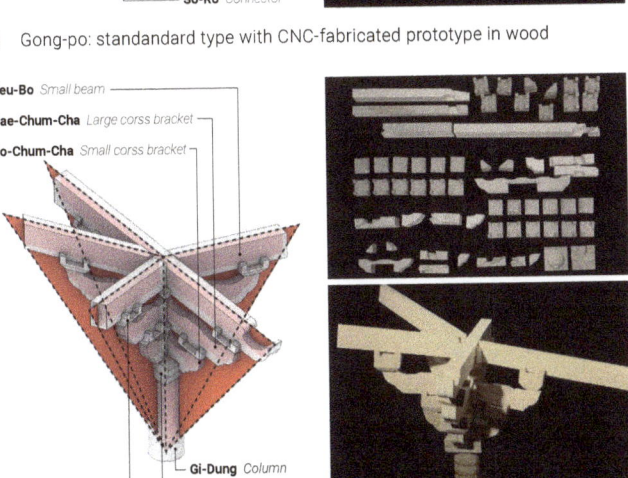

- **Teu-Bo** *Small beam*
- **Dae-Chum-Cha** *Large corss bracket*
- **So-Chum-Cha** *Small corss bracket*

- **Gi-Dung** *Column*
- **Ju-Du** *Main connector*
- **So-Ro** *Connector*

4   Gong-po: transformed type with CNC-fabricated prototype in wood

6   Detail in transformation (Seungwoo Kim, 2021)

the guide triangle, the overall shape is transformed and Booelean-ed parts are generated in a system of correlated information. They are fabricated by a 3-axis CNC machine and assembled by hand after CNC cutting (Figures 3, 4).

To deal with sophisticated and solid fabrication at a large scale, stereolithography (SLA, photocuring resin 3D printing) is employed. The machine for this project has 0.05 mm layer thickness with 0.01 mm accuracy by focusing an ultraviolet (UV) laser on to a vat of photopolymer resin. The computer generated form is divided into seven floors and fourteen parts to fit on a 1700 mm x 800 mm x 600 mm bed size. The parts are assembled on site with pre-designed and printed connectors, also inspired by Korean traditional joint systems without any metal connectors or adhesives.

CONCLUSION

In a recent survey in 2021 by the National Traditional House Center in Korea, 68.2% of local consumers responded that

they are willing to live at traditional Korean house. However, building technology and processes are still limited to the traditional production methods of a few skilled artisans. *Versatile Bracketry* can maximize the versatility of traditional

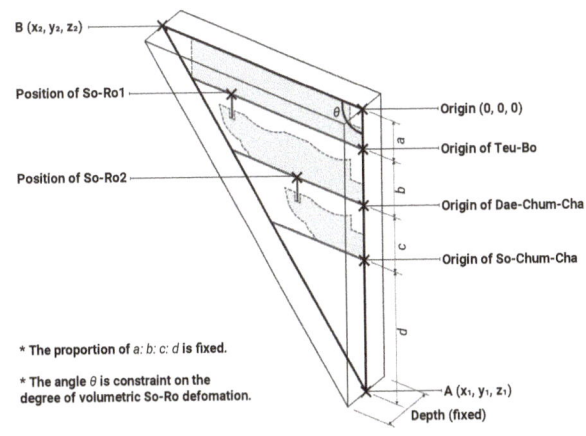

$B(x_2, y_2, z_2)$

Position of So-Ro1

Position of So-Ro2

Origin (0, 0, 0)

Origin of Teu-Bo

Origin of Dae-Chum-Cha

Origin of So-Chum-Cha

$A(x_1, y_1, z_1)$

Depth (fixed)

* The proportion of $a: b: c: d$ is fixed.

* The angle θ is constraint on the degree of volumetric So-Ro defomation.

7   The nodes in triangular system

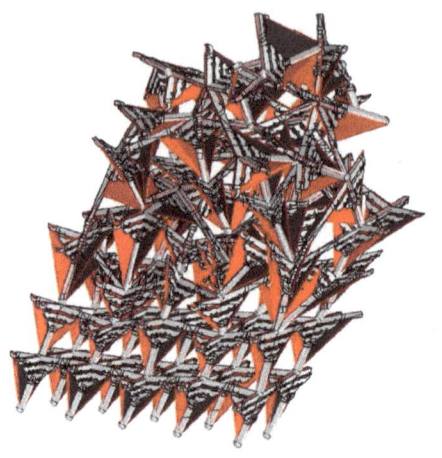

8 3D geometric network in triangular discipline.

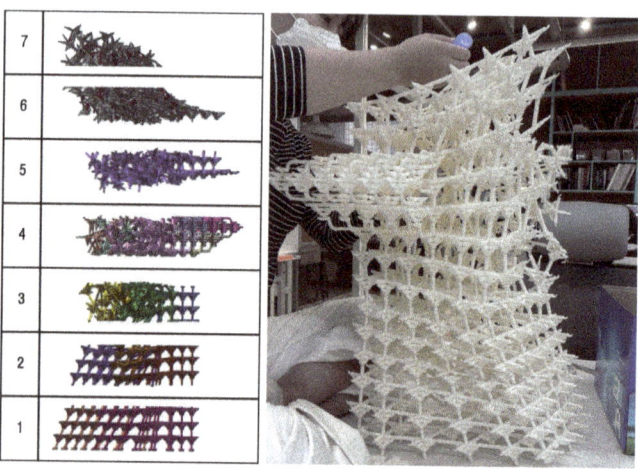

9 Divided parts to fit in bed size of stereolithography and scaled mock-up model

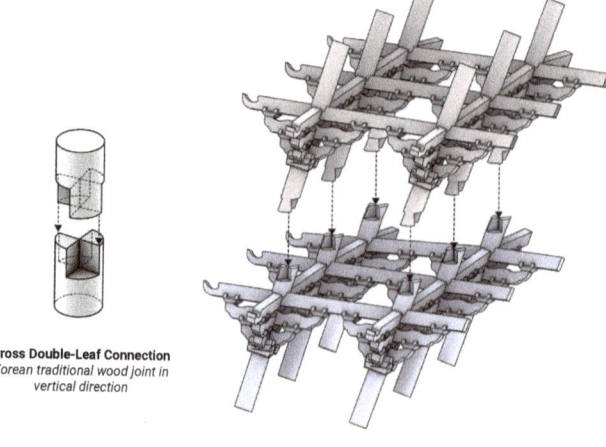

Cross Double-Leaf Connection
Korean traditional wood joint in vertical direction

10 Vertical assemby detail.

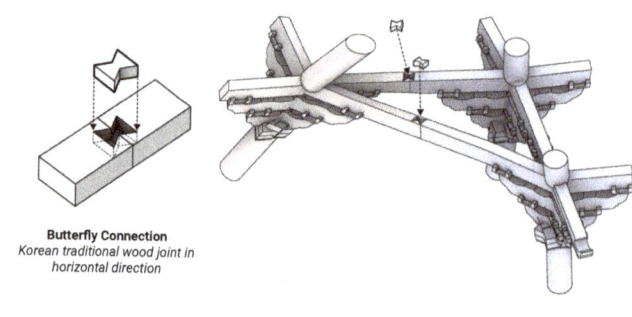

Butterfly Connection
Korean traditional wood joint in horizontal direction

11 Horizontal assembly detail

architecture to overcome the limitations of shape and material. It serves as an accelerator for utilizing modern computation technology in the architecture of contextualized design.

ACKNOWLEDGMENTS

This work is funded by the Korea Institute of Industrial Technology and supported by the Seoul National University of Science and Technology. I acknowledge the contribution of 3D Solution Corporate for prototyping and assembling. Special thanks to design team members Seohyeon Kwon, Jihyeon Ju, Seungwoo Kim, Yunah Park, and Jinhee Lee and to Seohui Yu for video editing. Finally, I thank everyone involved in the construction of the prototypes.

REFERENCES

Hwang, Jie-Eun. 2002. "A Study on the Analysis and Data Modeling for the Wooden Construction of Traditional Korean Architecture." *Journal of Architectural Institute of Korea* 18 (2): 82.

Leach, Neil and Philip F. Yuan. 2017. *Computational Design*. Shanghai: Tongji University Press.

Menges, Achim and Jan Knippers. 2021. *Architecture Research and Building ICD|ITke*. Basel: Birkhäuser.

Park, Jungdae. 2011. "A Study on the BIM-based Design for the Elements of Wooden Structure of Korean Traditional Buildings Through a Parametric Design Methodology" *Korean Journal of Computational Design and Engineering* 16 (2): 104–113.

Sabin, Jenny E. and Peter Lloyd Jones. 2018. *LabStudio: Design Research between Architecture and Biology*. New York: Routledge.

Shelden, Dennis. 2013. "Networked Space." In *Computational Works: The Building of Algorithmic Thought, AD, Architectural Design* 222, edited by Brady Peters and Xavier de Kesteller. Chichester: Wiley. 36–41.

Yoo, Youngjae. 1995. "Reforms in the Era of Globalization." *Journal of Public Administration in Gyeongsang National University* 7: 71–78.

12 Accumulated components (Seungwoo Kim, 2021)

13 Series of Gong-po

14 Gong-po Spectrum in Versatile Bracketry (Seungwoo Kim, 2021)

IMAGE CREDITS

Yong Ju Lee is an architect based in Seoul. He has worked on stimulating design for everyday life at multiple scales and media. His interest is geometric translation as a primary creative and aesthetic gesture of building, from conceptual process to actual construction. His works have been presented worldwide at institutions such as the Museum of Modern Art and Venice Biennale. He also received world-renowned awards including iF Design Award in 2020 and Design Vanguard from Architectural Record in 2021. He earned academic degrees from Yonsei University and Columbia University. He is currently Assistant Professor at Seoul National University of Science and Technology.

Convergence

Advancing Robotic Wire Arc Additive Manufacturing to the Architectural Scale in an Urban Context

Jenny E. Sabin, Michael Paraszczak, Dillon Pranger, John Hilla

Convergence celebrates the thriving, vibrant, and rich heritage of excellence of the University of Nebraska Medical Center through materialized concepts that embed change, transformation, and contemplation. The project incorporates the most advanced methods and innovations in digital and robotic fabrication with the integration of timeless and contextually sensitive materials that interact with the sun and human perception. The project features stainless steel wire arc additive manufacturing through robotic 3D printing, nonstandard CNC machined polycarbonate panels laminated with responsive wavelength-dependent dichroic film, and stainless-steel stiffener rings. Sited in the new Northwall Plaza, *Convergence* serves as the outdoor threshold to the buildings and the campus welcome center facilitating an ideal setting for conversations, fellowship, and engagement by students and faculty.

The largest 3D printed stainless steel sculpture in the world, *Convergence* builds on the expanding field of architectural additive manufacturing through a unique collaboration with Lincoln Electric by pushing the boundaries of architectural scale seen in the wire arc additive manufacturing (WAAM) process. The WAAM fabrication process was previously brought into the architectural space by ARUP in their research into optimizing metal structural building elements (Galjaard et al. 2015). The design opportunities of this emerging method were showcased in the "MX3D" Bridge project designed and developed by the Joris Laarman Lab and MX3D (Laarman 2018). Designers are just starting to explore this new opportunity for melding the versatility and efficiency of 3D printing with the strength of metallic materials.

Roland Snooks expanded on this new method in "Remnants of a Future Architecture" exhibited at Melbourne Design Week 2022 (Snooks 2022). At 28 feet in height and 12 feet in diameter, *Convergence* acts as a beacon in the UnMC plaza and pushes WAAM to the architectural scale in an urban university context. Through the unique WAAM process, the structure—at 4,000 lbs of robotically 3D

PRODUCTION NOTES

Architect:	Jenny Sabin Studio
Client:	University of Nebraska Medical Center and the Nebraska Arts Council, 1% for Art Program
Status:	Built
Site Area:	28' height x 12' diameter
Location:	Omaha, Nebraska
Date:	2021

1   Opening of *Convergence* at UNMC Northwall Plaza (Jenny Sabin Studio, 2021)

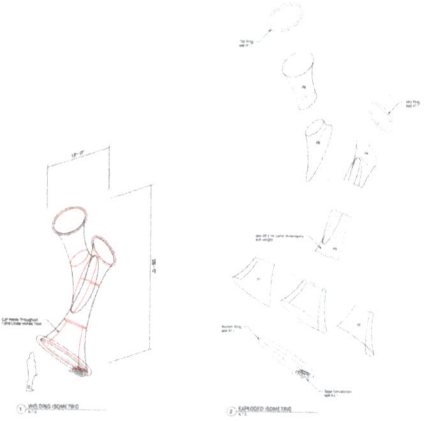

2 Fabrication planning diagram: the central spine was split into nine total parts to reduce overhang, thermal distortion, and print error

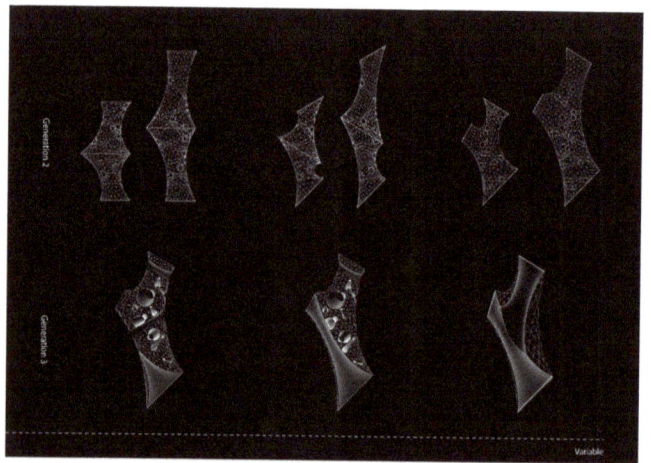

3 To amplify the spatial presence of the project through equal and opposite forces, the interactive form-finding process included anchoring and balancing the structure on a single point that meets the foundation in the plaza

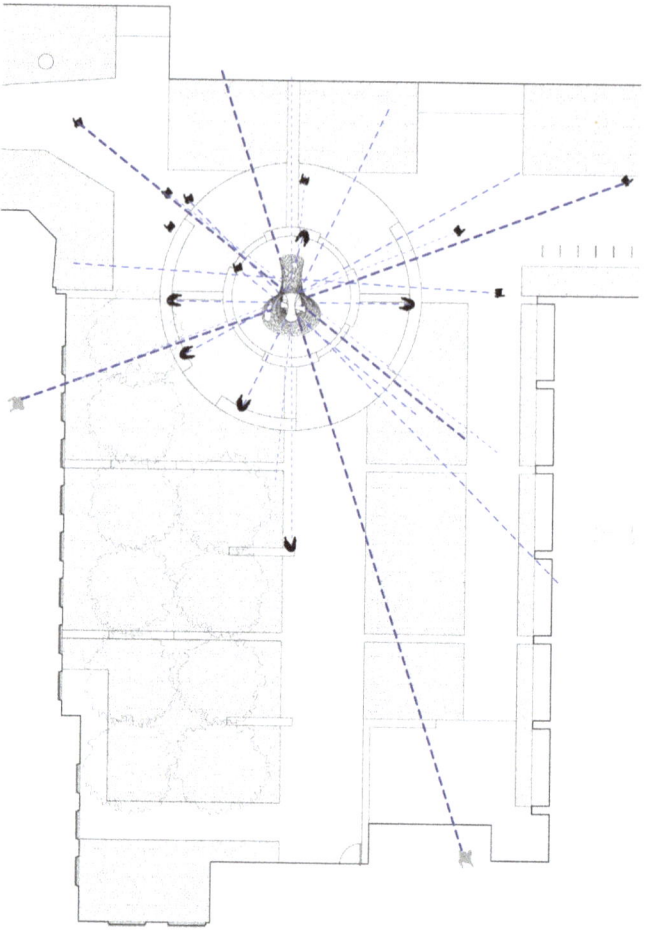

4 Generative design process, including site and view analysis of the plaza; voids within the project were calibrated and optimized relative to views from adjacent public balconies and from the plaza grounds looking through the structure

printed stainless steel material organized in nine unique printed panels—uses less steel material than traditional fabrication methods, thus reducing waste and saving time and labor.

The design process for *Convergence* commenced with a comprehensive site analysis of the local plaza and its connections to other permanent works of public art on campus. Our form-finding process was developed with custom scripts in Kangaroo, an interactive real-time physics engine for simulating dynamic forces. Voids within the project were calibrated, iterated, and optimized relative to views from adjacent public balconies and from the plaza grounds looking up and through the structure. Working closely with Lincoln Electric, fabrication constraints, such as scale of parts and maximum draft angle or overhang within the panels, were integrated

and optimized. For example, during two test prints at Lincoln Electric, 3D scans of the finished parts revealed substantial thermal distortion at some locations of the part whereas others nearly matched the CAD model. After further tests, it was determined that a print deviation of less than 0.125 inch was acceptable from an engineering perspective for final fabrication and welding fit-up of all parts into the singular spine structure. Further, due to the steep draft angles, extra material—such as flat panels—was added to the CAD model to stabilize the part during the printing process. These areas were then removed during the finishing phase. The fabrication strategy used a single welding process, stacked vertically in an additive method. Specifically, BLUE MAX® MIG 316LSI stainless steel was used to additively manufacture each of the nine 0.25" thick parts via Wire Directed Energy Deposition (Wire DED) on a

Convergence Sabin, Paraszczak, Pranger, Hilla

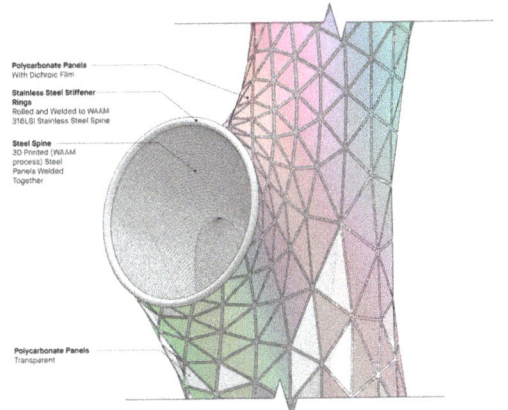

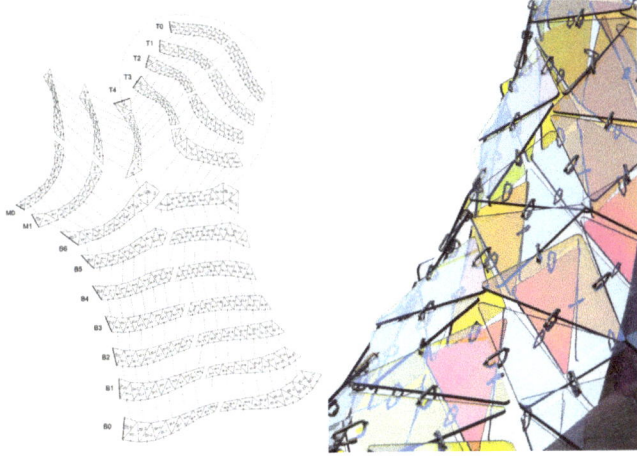

5　Material and structural diagram highlighting the polycarbonate dichroic mesh, WAAM steel spine, and stainless steel ring stiffeners

7　To fabricate the panels, 0.75" polycarbonate flat sheets were laminated with wavelength dependent dichroic film and cut via CNC process

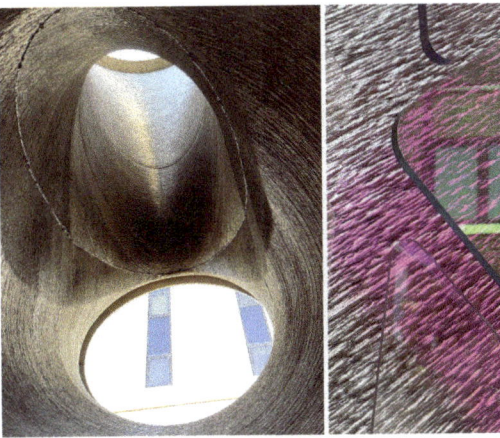

6　Section C of the project being printed in a cell at Lincoln Electric; BLUE MAX® MIG 316LSI was used to additively manufacture each of the nine 0.25" thick parts via Wire Directed Energy Deposition (Wire DED) on a 0.5" thick ASTM A36 base plate; the method of DED was GMAW-Pulse

8　Material details of *Convergence*: interior of the 3D printed stainless steel spine (left); detail of the exterior of the steel structure featuring a texture of continuous layers of WAAM steel and dichroic mesh panels (right)

0.5" thick ASTM A36 base plate. This base plate was later removed during the final fabrication and weld fit-up of the spine structure. The material was selected after two cylindrical tests were produced by Lincoln Electric. Initially, attempts were made to remove the presence of visible arc start locations through grinding and wire brushing, but it was determined that this would create noticeable grind patches; due to labor, cost, and time, the ribbed texture with periodic drips or bumps where the welding arc starts was preferred as it provides a beautiful texture and narrative of the project's robotic wire arc production. In collaboration with Lincoln Electric and during the tool path planning process, it was determined that due to the size of the prints—the largest measuring at 7 feet tall—the start and stop locations of the fabrication process had to be situated on the side of the part closest to the robot. Once

the printing process was completed at Lincoln Electric, all parts were shipped to our fabricator in Omaha for final fabrication, welding, and finishing.

During the robotic fabrication testing phase and design iterations of the spine, form-finding experiments were conducted to generate the exterior tessellated mesh composed of clear and dichroic triangular components. The surface was treated as a fabric membrane and simulated in Kangaroo with anchors situated at the three ring locations of the central spine. These digital outputs were then tested at model scale through physical prototypes. Finally, in collaboration with our engineers, a strategy was developed and scripted to place and coordinate holes in each panel to stitch and connect the mesh assembly together. Strips of the mesh were assembled with stainless steel carabiners and flat packed for shipment to Omaha, Nebraska.

Hybrids & Haecceities

9 Attempts were made to remove the presence of visible arc start locations through grinding, but it was determined that this would create noticeable grind patches; due to labor, cost, and time, the ribbed texture with periodic bumps provided a beautiful texture and narrative of its robotic wire arc production

10 *Convergence* at night casting dynamic color and light; gradient studies were scripted to test variations in density of the dichroic and clear polycarbonate panels to optimize and highlight reflected and refracted light as washes of color and shadow during the day and night

As a set of integrated material relationships, the design for *Convergence* incorporates part-to-whole 3D printed and machined component strategies to realize a project that celebrates lightness, dynamic structure, and equilibrium through equal and opposite forces. Taking inspiration from the University of Nebraska Medical Center's pivotal role in global health through their response and leadership in both the Ebola epidemic and now the Coronavirus pandemic, *Convergence* celebrates people coming together from around the world. Through part-to-whole, individual, and collective, the project reflects and celebrates the faculty, doctors, students, researchers, administrators, and community champions who have left their mark and defined UNMC's past, present, and future.

ACKNOWLEDGMENTS

Engineering Design: Powell Draper, Matthias Peltz, SBP New York City; Wire Arc Additive Manufacturing: Lincoln Electric Additive Solutions; Fabricators and Installers: Prairie Mechanical Corp., Patriot Service Inc., Accufab; Commissioned by University of Nebraska Medical Center & the Nebraska Arts Council 1% for Art Program; Special thanks to Dr. Northwall and the Northwall family and Chancellor Gold for supporting this project.

REFERENCES

Galjaard, S., S. Hofman, N. Perry, and R. Shibo. 2015. "Optimizing Structural Building Elements in Metal by using Additive Manufacturing" In *IASS Symposium 2015: Future Visions; Proceedings of International Association of Shell and Spatial Structures*. Amsterdam, Netherlands.

Laarman, Joris. 2018. "MX3D Bridge." Amsterdam, Netherlands: Dutch Design Week.

Snooks, Roland. 2022. "Remnants of Future Architecture." Melbourne, Australia: Melbourne Design Week.

IMAGE CREDITS

Figures 1-5, 7-8, 10-11 : ©Jenny Sabin Studio, 2021
Figures 6, 11: ©Lincoln Electric Company, 2021
All other drawings and images by the authors

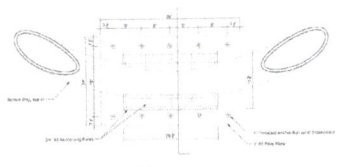

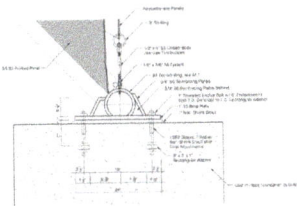

11 After the spine was assembled and welded together, the 28-foot structure was passivated and wire-brushed to remove residue from the printing process, protect the stainless steel from the outdoor elements, and enhance the luster of the steel material. Due to constraints of the site, a crane rigged and hoisted the structure over a sky bridge before setting and mounting it to its foundation. After the spine was bolted in place, the strips of the polycarbonate surface were installed concentrically and connected with stainless steel carabiners. Finally, turnbuckles were added at each stainless-steel ring termination allowing for tension adjustments to the surface.

Jenny E. Sabin is the Arthur L. and Isabel B. Wiesenberger Professor in Architecture and Associate Dean for Design at Cornell College of Architecture, Art, and Planning where she established a new advanced research degree in Matter Design Computation. She is principal of Jenny Sabin Studio, an experimental architectural design studio based in Ithaca and Director of the Sabin Lab at Cornell AAP. In 2017, Sabin won MoMA & MoMA PS1's Young Architects Program with her submission *Lumen*.

Michael Paraszczak holds a MArch from Cornell University and is a Junior Designer with Jenny Sabin Studio. Paraszczak holds a BS in Architecture from SUNY University at Buffalo and has additional experience working in the design office of REX and as technician at the University at Buffalo Fabrication Workshop.

Dillon Pranger is a licensed architect in the State of Illinois and is Associate Architect with Jenny Sabin Studio. He holds a BS in Architecture from the University of Cincinnati, and a MArch from Cornell University. Pranger has taught extensively at institutions such as Cornell AAP, Harvard GSD, and Syracuse University SOA, with additional experience working in the design offices of KPF and Eisenman Architects.

John Hilla is a Senior Designer with Jenny Sabin Studio, a Research Associate with JSLab, and a recipient of the American Institute of Architects Henry Adams Medal. Hilla holds a MArch from the University of Pennsylvania, as well as a BS in Mechanical Engineering and a MS in Engineering Management from Syracuse University.

Nurse Pod

A Critical Examination of the Role of Decay in Architecture

Marc Swackhamer, Blair Satterfield, Brian Buma, Hadley Rhodes, Matthew Hayes, Logan Ebert, Jon Ackerley

THE PROBLEM

Historically, fires were key to forest lifecycles; post-fire regeneration created the next generation of trees. Today, however, fires are burning hotter and spreading faster than previously recorded, breaking the rhythm of the traditional regrowth cycle.

In 2020 alone, Colorado saw three of the largest wildfires recorded in state history. The intensity of these fires left native soils sterile and increased soil erosion, damaging the landscape and watershed. Despite human efforts to replant lost forests in the face of a warming and drying climate, newly planted saplings often experience a high early mortality rate due to soil loss, sterilization, and low water availability. *Nurse Pod* aims to assist in forest regeneration by providing a supportive system for vulnerable tree saplings.

THE CONCEPT

In 2020, we formed the multidisciplinary, multi-institutional *Nurse Pod* team to develop prototypes informed by ecology, architecture, art, and engineering. This collaboration generated a design process that critically considers the notion of life support, decay, and the relationships between a designed object, its surroundings, and its purpose. By

PRODUCTION NOTES

Renderings:	Rhino, Twinmotion, Photoshop
Diagrams:	Rhino, Illustrator
Prototypes:	Developed in Rhino/Grasshopper. Printed on Prusa MK3S with crushed stone/PLA composite filament.

1 *Nurse Pod* rendering

3 Pine Gulch fire

4 Pine Gulch fire

5 *Nurse Pod* rendered view

putting a design's lifecycle at the forefront of its development, we can better understand how our projects are wed to the environments in which they are built.

THE DESIGN CRITERIA

The established design criteria for the *Nurse Pod* are a system that collects, retains, and distributes resources to a sapling tree while dissolving into its surrounding environment, matching its pace of decay with the growth of the new tree. Inspiration for the initial concept was derived from the benefits of naturally occurring nurse logs, fallen trees that elevate and protect young seedlings in established forests. As they decompose, the pods sequester and distribute water and provide key nutrients for adolescent plants. We tested initial concept iterations in controlled greenhouse experiments, yielding results that informed the current prototype.

In collaboration with a forestry biologist, and learning from our previous experiments, the following criteria guided the most recent prototype:

1. Nurse and support a tree sapling or seed

2. Collect and sequester water for prolonged dry events

3. Manage the rate at which water is allocated to the sapling/seed

4. Use of a benign material that biodegrades over time (PLA composite, paper pulp, wood pulp, clay)

5. Ease of transfer, assembly and deployment

CONCEPT IMPLEMENTATION

With a focus on fire-stricken forests, we split the original monolithic design into two separate modules, allowing us to separate water retention from sapling regrowth. The form of *Nurse Pod* maximizes water intake and facilitates the traversal of sapling roots to the soil below.

Nurse Pod Swackhamer, Satterfield, Buma, Rhodes, Hayes, Ebert, Ackerley

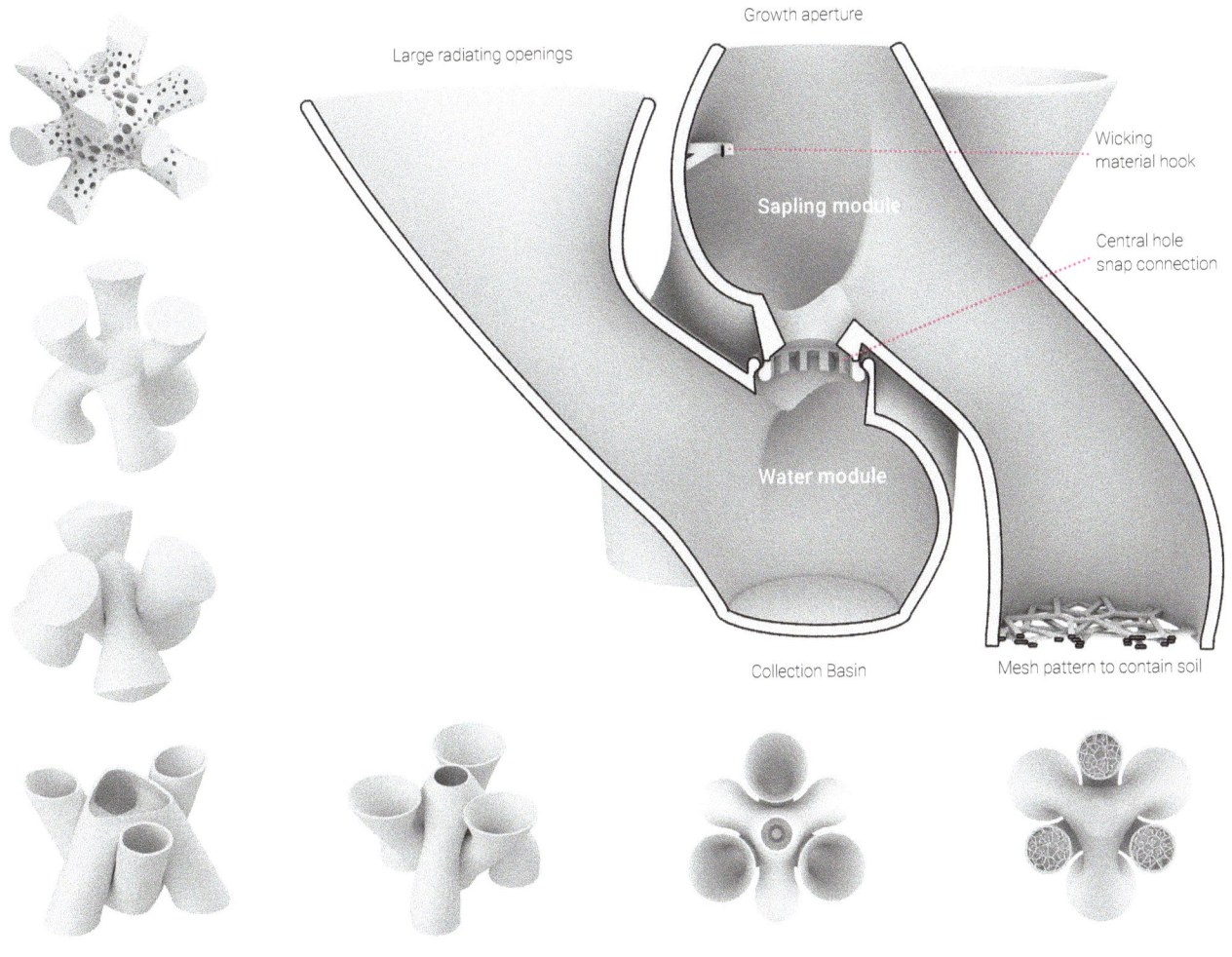

Large radiating openings

Growth aperture

Wicking material hook

Sapling module

Central hole snap connection

Water module

Collection Basin

Mesh pattern to contain soil

6 Early design iterations

7 *Nurse Pod* section with top and bottom views

The modules connect at a centralized aperture, allowing wicking material to regulate water distribution into the soil module. We printed the current prototype from biodegradable PLA (Polylactic Acid, a renewable bioplastic) and crushed stone composite material to allow it to naturally decay over time. Future designs will test other printable biomaterials, such as recycled wood pulp and paper pulp or locally sourced clay.

Through a new series of controlled greenhouse experiments in summer of 2022, we will collect data on water volume, retention, and distribution relative to the sapling's growth. This data will inform the next generation of *Nurse Pod* prototypes.

THE WHY

Nurse Pod serves as a critique of the way architects think about our built work in relationship to the natural world.

By carefully considering the full lifecycle of an object, or building, from its construction to its eventual decay, the question of material as contaminant versus material as nutrient becomes central. Perhaps eventually we will consider buildings as contributors to the biological cycles of growth and decay that they often fail to acknowledge.

Nurse Pod's approach to addressing a major global challenge—here, the loss of forests due to wildfire and climate change—models the kind of unlikely solution that grows from a deep investment in post-disciplinary collaboration and its associated research, which is necessarily slow and resistant to entrenched, discipline-specific approaches.

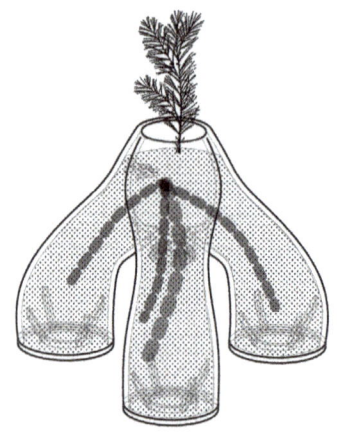

Soil Module is packed with nutrient rich soil and wicking material strands that reach up through the center into each leg

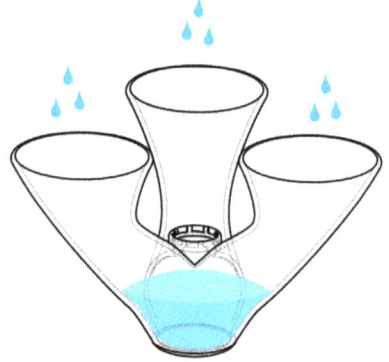

Water Module has and large central connection and large radiating openings for water collection

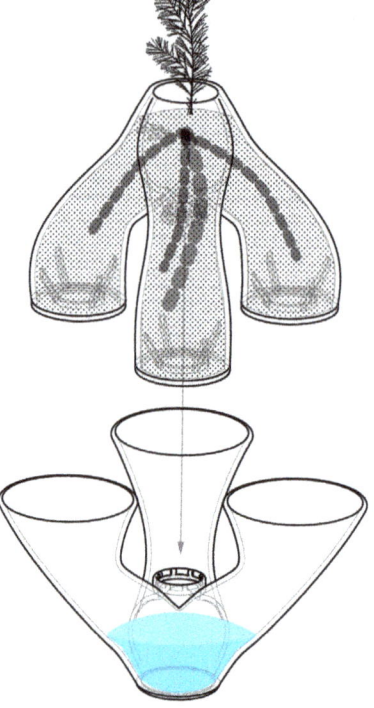

The two modules snap together at the center, allowing the wicking material to drop into the water module

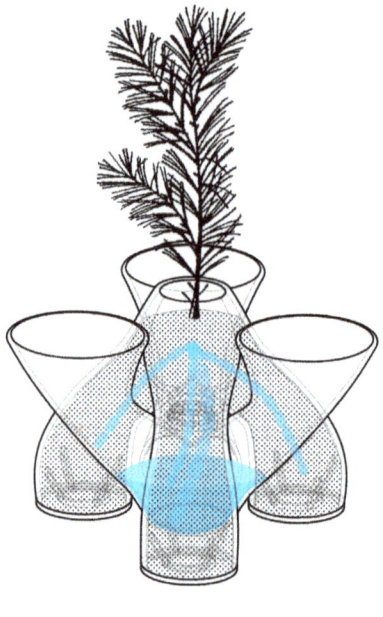

When the two modules are nested together, the wicking material steadily moves water vertically into the soil module

8   Assembly and resource management series diagram

ACKNOWLEDGMENTS

We would like to acknowledge support and resources from the following institutions:

The University of Colorado Denver, College of Architecture and Planning and College of Liberal Arts & Sciences Deparment of Integrative Biology

The Univeristy of British Colombia, School of Architecture & Hilo Lab

REFERENCES

Buma, B., S. Weiss, K. Hayes, and M. Lucash. 2020. "Wildland fire reburning trends across the US West suggest only short-term negative feedback and differing climatic effects." *Environmental Research Letters* 15 (3): 034026.

Grand Junction Field Office, Bureau of Land Management. "Pine Gulch Fire - Photographs." InciWeb the Incident Information System - Pine Gulch Fire Photographs. Accessed July 3, 2022. https://inciweb.nwcg.gov/incident/photographs/6906/0/.

Rother, M.T., and T.T. Veblen. 2017. "Climate drives episodic conifer establishment after fire in dry ponderosa pine forests of the Colorado Front Range, USA." *Forests* 8(5): 159.

Rother, M.T., and T.T. Veblen. 2016. "Limited conifer regeneration following wildfires in dry ponderosa pine forests of the Colorado Front Range." *Ecosphere* 7(12): e01594.

9 *Nurse Pod* in greenhouse (LoDo Lab, July 2022)

10 Greenhouse Experiment (LoDo Lab, July 2022)

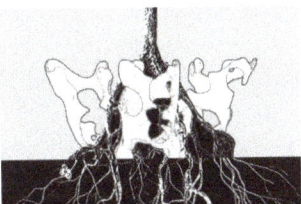

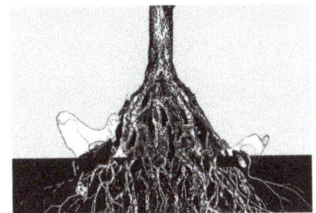

11 Speculative drawing of *Nurse Pod* decay and sapling growth over time

Marc Swackhamer is Department Chair and Professor of Architecture at the University of Colorado Denver, College of Architecture and Planning. He is also the director and founder of LoDo Lab and co-founder of HouMinn Practice.

Blair Satterfield is the Chair and Associate Professor of Architecture at the University of British Columbia, School of Architecture and Landscape Architecture. He is also the director and founder of HiLo Lab and co-founder of HouMinn Practice.

Brian Buma, PhD is Associate Professor in the Deparment of Integrative Biology at the University of Colorado Denver. He is the graduate program director and has been a member of LoDo Lab since 2019.

Matthew Hayes is a research assistant with LoDo Lab and a current MArch candidate at University of Colorado Denver's College of Architecture and Planning. He holds an AA and BS in Architecture degrees with minors in Landscape Architecture and City & Regional Planning from The Ohio State University.

Jacob Taswell is a research assistant with LoDo Lab and a current MArch candidate at University of Colorado Denver's College of Architecture and Planning. He holds a BA in Music from Yale University.

Hadley Rhodes is a research assistant with LoDo Lab and a current MArch candidate at University of Colorado Denver's College of Architecture and Planning. She holds a Bachelor of Science from the Denver School of Nursing.

Logan Ebert is a research assistant with LoDo Lab and a current MArch candidate at University of Colorado Denver's College of Architecture and Planning. He holds a BA in Emergent Digital Practices from University of Denver.

Jon Ackerley is a research assistant with HiLo Lab and a recent MArch graduate of the University of British Columbia, School of Architecture and Landscape Architecture.

ONDA Wall

Using Patterns to Fuse Topology with Topography

Kory Bieg

The *ONDA Wall* (Figure 2) was designed with two intentions: as a didactic tool for teaching digital design and fabrication, and as an exploration in contemporary architectural theory, specifically how architecture might manifest some ideas held true by the philosophy of Object-Oriented Ontology (OOO).

TECHNIQUE AND MATERIALITY

As a tool for teaching, the design of the project included the use of multiple software programs so that future students could see and understand how and why one might use one program over another, and more importantly, how they can be used concurrently to take advantage of the design features of each. For *ONDA Wall*, students used Autodesk 3ds Max, Rhinoceros with Grasshopper, and Processing 3d.

The project is a series of over five thousand unique Computer Numerical Controlled (CNC) fabricated parts (Figure 2), half of which are painted Medium-Density Fiberboard (MDF), while the other half are translucent High-Density Polyethylene (HDPE). The profile, or form, of the painted MDF parts is determined by a procedurally generated topography developed through an iterative process of trial and error using Autodesk 3ds Max.

The topography of the HDPE parts was determined by a secondary surface derived from the contours of The University of Texas at Austin campus. The two topographies weave in and out as they attempt to balance and tie together the other effects present in the project.

PRODUCTION NOTES

Architect: Kory Bieg

Status: Built

Location: The University of Texas at Austin School of Architecture

Date: 2021

1 Left oblique photograph showing anamorphic projection of rectangular figure

2 Elevation photograph

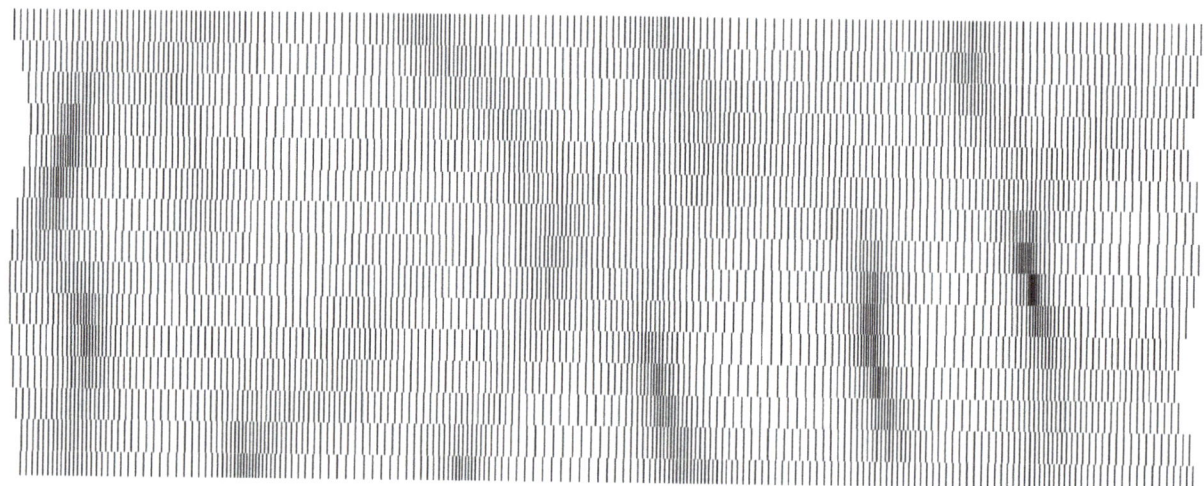

3 Perlin noise pattern map

Processing 3d, a scripting program used by artists, architects, and other creatives, was used to determine the spacing between parts. We used Perlin noise (Figure 3), a random value generator that simulates natural processes and patterns. Looking directly at the wall, one sees a set of organic figures emerging from the shadows between the MDF and HDPE parts (Figure 5). These figures are independent from the two topographies used to develop the profile of the parts. The spacing pattern provides a third, ephemeral effect that binds with the other two, especially when the walls backlighting is turned on. Finally, we used Rhinoceros and Grasshopper, the node-based visual programmer, to finish detail, fabrication, and assembly files for the project (Figures 4, 6, 7).

DESIGN THEORY

The project is an exploration in the application of OOO as an architectural design pedagogy by subverting hierarchy at every scale. Though there are multiple overlapping geometries and effects that come through in the design, no one condition is primary. One tenet of OOO is that all objects are equal, whether they be oranges, antelope, or handkerchiefs (Bryant 2011). In the same vein, no effect overwhelms, or fully yields to, another. However, this does not mean that all effects are expressed equally at all times.

Another important tenet of OOO, is that objects have properties that are withdrawn. One can never fully know an object; there are always some qualities held in reserve (Harman 2018).

4   Individual parts grouped into modular assemblies

5   Detail photograph

Similarly, *ONDA Wall* is packed with qualities, effects, form, and pattern, of which only a portion can be seen at any one time. In *ONDA Wall*, form and effect emerge depending on when and from where one views the wall. If one looks at the wall at an angle from the left side, one will see not only a reflection of the painted MDF color, but also an anamorphically projected, cyan colored rectangle (Figure 1).

When viewed from the right side, one will see a gradient of color and only a faint cloud of cyan as it reflects off the HPDE (Figure 8). Both effects are captured from straight on, but the pattern formed by the irregular spacing between parts is brought forward. Although there are no moving parts, the wall is incredibly dynamic and encourages interactivity.

PROJECT LOCATION

The project is located on a newly renovated floor of the West Mall Building on the The University of Texas at Austin campus. The renovation includes three new digitally enabled classrooms that were specifically designed for teaching visual communication courses to undergraduate and graduate students in the School of Architecture. The wall serves as a demonstration of what can be produced using some of the tools they are learning. As new tools find their way into the curriculum, the wall will be replaced with a new project highlighting the use of these emerging tools, encouraging interactivity.

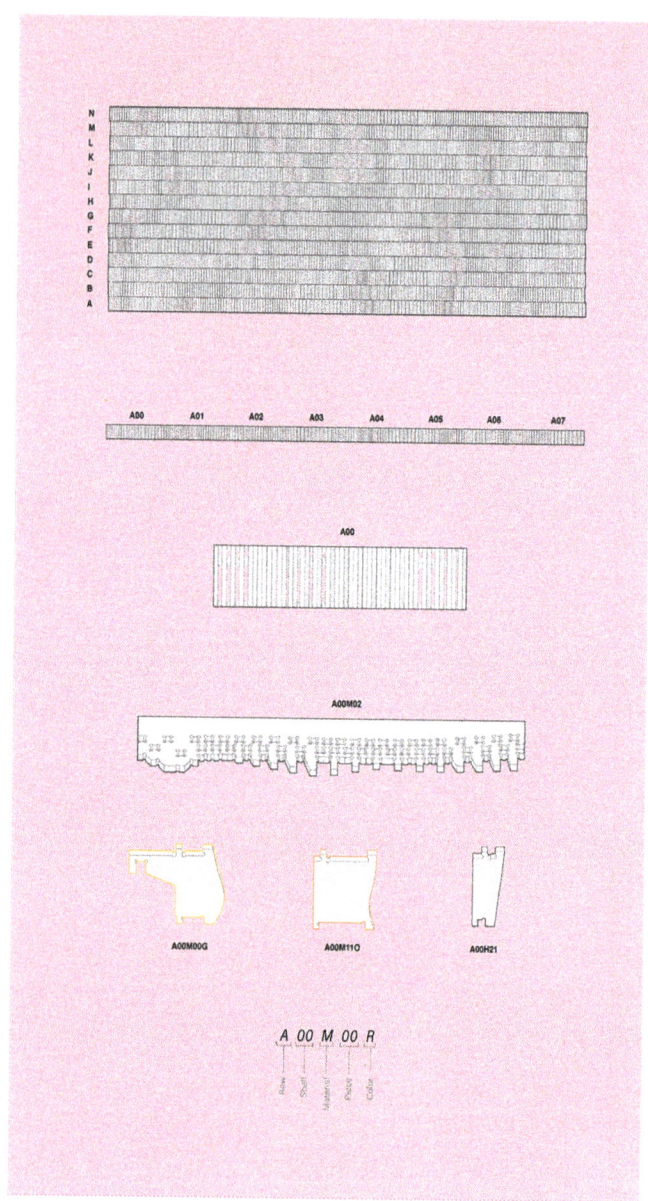

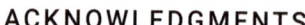

6   Elevation of pattern and part naming guide

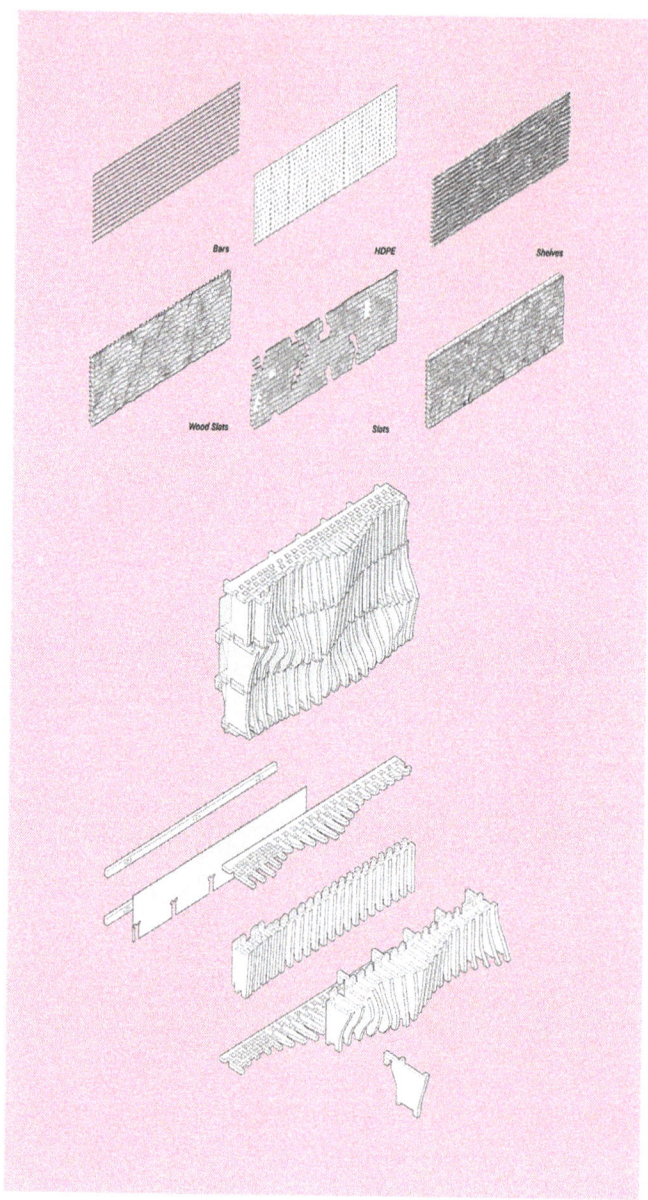

7   Exploded axonometric of layers and one assembled unit

ACKNOWLEDGMENTS

This installation was made possible by the generous support of The University of Texas at Austin School of Architecture and Dean Michelle Addington.

Initial design and research for *ONDA Wall* was completed in a studio taught by Kory Bieg at The University of Texas at Austin School of Architecture. The project team included Abby Abolt, Bruno Canales, Jayme Greene, Marianna Jones, Shelly Kimmel, Danielle Ndubisi, Erik Olivarez, Davis Richardson, Sam Shiminski, Sarah Spielman, Caroline Stacey, Joel Sterling, Raquel Valdez, and Cole Wendling.

REFERENCES

Bryant, Levi R. 2011. *The Democracy of Objects*. Ann Arbor: Open Humanities Press.

Harman, Graham. 2018. *Object-Oriented Ontology: A New Theory of Everything*. Great Britain: Pelican Books.

IMAGE CREDITS

Figures 3-4, 6-7: Abby Abolt, Bruno Canales, Jayme Greene, Marianna Jones, Erik Olivarez, Davis Richardson, Raquel Valdez.

All other images by the author.

8 Right oblique photograph showing color gradation and overlapping topographies

Kory Bieg is Associate Professor of Architecture and Program Director for Architecture at The University of Texas at Austin. He received his Master of Architecture from Columbia University and is a registered architect in the state of Texas. Since 2013, he has served as Chair of the TxA Emerging Design + Technology conference, and co-Director of TEX-FAB Digital Fabrication Alliance. He has served on the Board of SXSW Eco Place by Design and the Association for Computer Aided Design in Architecture.

UNDERSTOREY

A Pavilion in Parts

Viola Ago, Hans Tursack

In the summer of 2018, our collaboration was awarded a University Design Fellowship from the Exhibit Columbus organization to design, fabricate, and build a large pavilion in Columbus, Indiana as part of a biannual contemporary architecture exhibition. Our proposal for the competition was a pavilion that would double as an ecological education center. Our inspiration for this program was triggered in part by our reading of Jane Bennett's materialist philosophy outlined in her book *Vibrant Matter* (2009). Through Bennett's lens, our design rendered our site's context as an animate field, replete with pre-existing material composites that we wanted to celebrate through a series of displays, information boards, and artificial lighting. In this, the installation would feature samples of local plants, minerals, and rocks, indigenous to Southern Indiana.

PRODUCTION NOTES

Architect: Viola Ago, Hans Tursack

Client: Exhibit Columbus, Landmark Columbus Foundation

Status: Built

Site Area: 20' length, 15' width, 14' height

Location: Columbus, Indiana

Date: 2019

1   View of pavilion on site

2 View of pavilion interior

4 Assembling on site with unskilled labor

3 View of ecological and geological samples, low display tables, pavilion interior

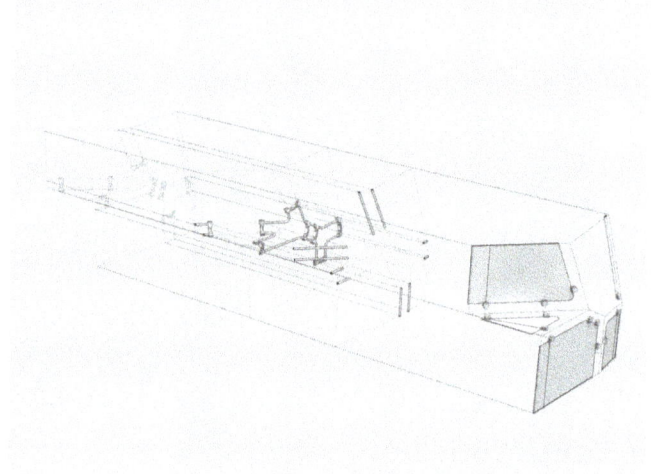

5 Exploded diagram; detail of structural connections and cladding

The pavilion design borrowed aesthetic cues from early British High-Tech architecture (Richard Rogers), ecological artists, sculptural precedents (Monika Sosnowska and Mateusz Von Motz), and from vernacular greenhouse construction; its cladding was semi-translucent and was conceived as a "kit-of-parts" to be assembled on site in a matter of a few days with unskilled labor.

The irregular geometry of the pavilion was originally imagined as a solid volume. As we resolved the project's tectonic character, however, the pavilion began to be read as a much lighter assembly of folded surfaces wedded to a wireframe skeleton. The frame was fabricated at the Autodesk Technology Center in South Boston and assembled on site. Our design exclusively used mechanical connections, standard hardware (it required no welding), and site-friendly earth anchors in place of concrete

foundations. A Karamba simulation allowed us to analyze and adjust the frame during the design phase. The final dimensions of the structure measured 20' length, 15 width, 14' height.

The structure of the pavilion was designed as a series of straight and bent steel tubes. We outsourced the production of the bent tube elements to a precision tube-bending fabrication company in Canada; the fabricated tubes were later powder coated with a protective yellow finish. The cladding for our pavilion was donated by Kalwall High Performance Translucent Building Systems in collaboration with the Parabeam 3D Fabric company. The material is a translucent, 5/16" inch Fiber-Reinforced Resin composite. The irregular geometry of the pavilion required us to water-jet the cladding material into 34 unique panel shapes. The most critical component of the pavilion's design was a repeating milled block aluminum 3-way

Understorey Ago, Tursack

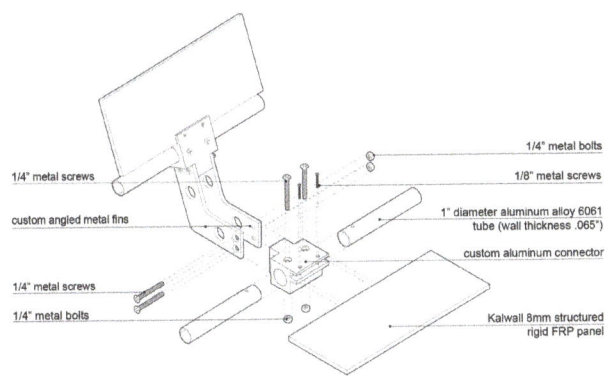

EXPLODED CONNECTION DETAIL
.25 : 1

1/4" metal bolts
1/8" metal screws
1" diameter aluminum alloy 6061 tube (wall thickness .065")
custom aluminum connector
Kalwall 8mm structured rigid FRP panel

1/4" metal screws
custom angled metal fins
1/4" metal screws
1/4" metal bolts

6 Exploded axonometric of typical connection detail

7 Closeup view of the panel to panel connections: accommodating for all unique angles with 2D-cut parts

8 First connection prototype

joint that acted as a universal connector. The joint negotiated multiple connections where the cladding, steel tubes, and 1/4" waterjet aluminum "fins" intersected with one another at 144 points.

The entire structure was fabricated without a printed drawing set. We used a highly detailed digital model generated in Rhino 3D based on the Grasshopper Visual Programming language, to generate files for 3D printers, a CNC Precision metal mill, a large-size CNC waterjet, and our tube-bending contractor. To avoid welding, on-site cutting, concrete pouring, and other processes that are contingent upon the skill of the fabricator, we modeled and fabricated the pavilion down to 1/32" coefficient of error.

While our pavilion was semi-temporary, we understand its design and construction as a prototype for future building-scale applications. The promise of our study is that our files could be sent to fabricators anywhere in the world to reproduce components such as our universal, aluminum-milled joint. The pavilion suggests flexible re-applications of the joint, steel tube, and FRP panel system at a variety of scales and sites with mail-order hardware, easily reproducible CNC components, and unskilled labor. Though quickly assembled, the project withstood the volatile climate of Southern Indiana during its 9+ month lifespan, and it was disassembled into its reusable component parts in a matter of days by two contractors with simple hand tools.

ACKNOWLEDGMENTS

Design Team: Hannah Daugherty, David Alcala, Lisa Kuhn, Aaron Payne

Fabrication Team: Taylor Boes, Zach Schumacher

Installation and Assembly: Megan Pettner, Becca Schalip

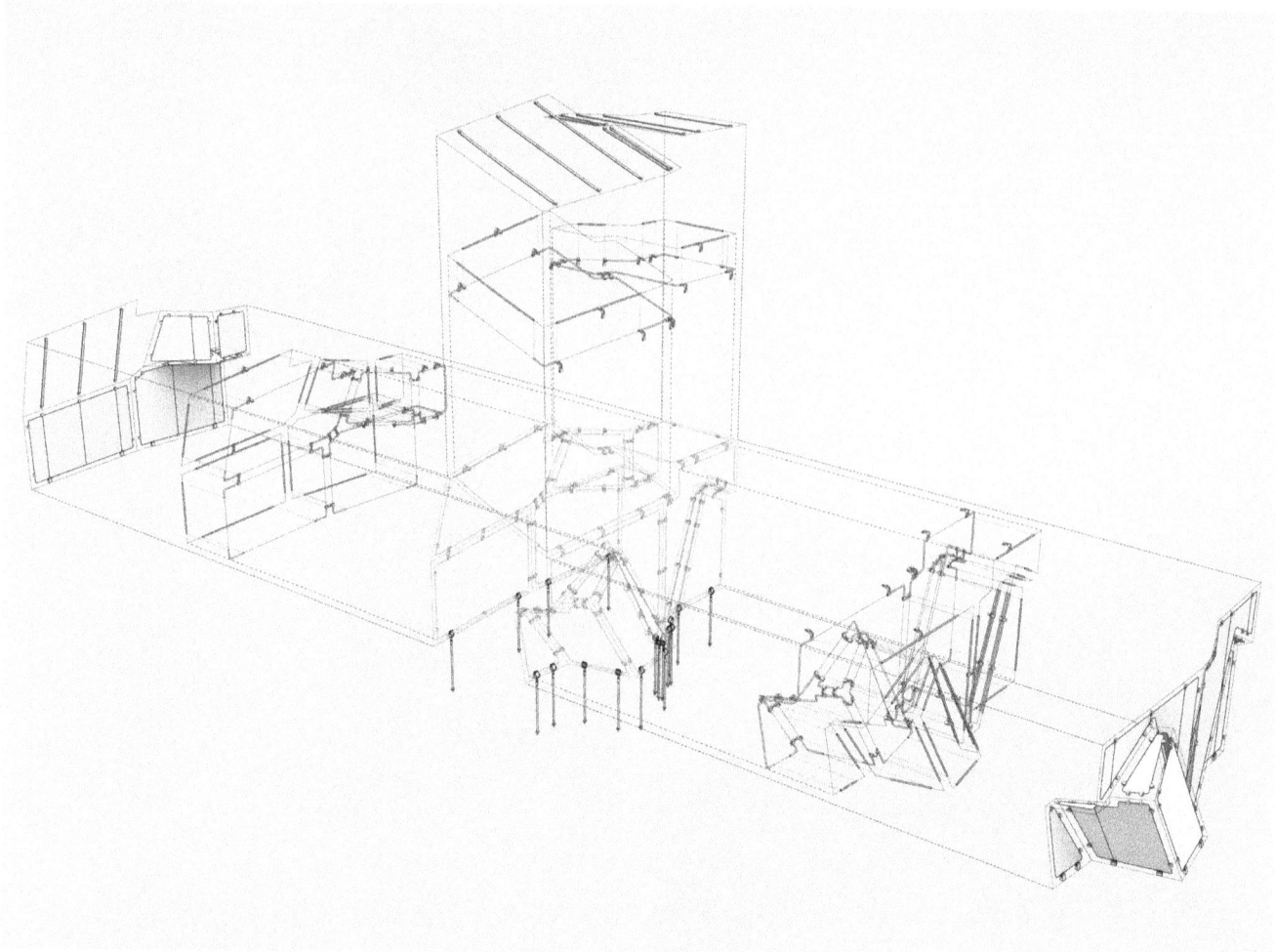

9 Exploded diagram of assembly of parts

Sponsorship: Ohio State University Knowlton School of Architecture; MIT School of Architecture + Planning; The Council for the Arts at MIT (CAMIT); Autodesk BUILD Space; Kalwall High Performance Translucent Building Systems; and Parabeam 3D Fabric Company.

IMAGE CREDITS

All photographs and other drawings and images by the authors.

Viola Ago (b. Lushnjë, Albania) is an architectural designer, educator, and practitioner. She directs MIRACLES Architecture. Recently, Viola held the Yessios Visiting Professorship at the Knowlton School of Architecture at OSU, the Muschenheim Fellowship position at the Taubman College of Architecture, University of Michigan, and the Wortham Fellowship at the Rice School of Architecture. Viola earned her M.Arch degree from SCI-Arc, and a B.ArchSci from Ryerson University. Her written work has been published in *Log*, Wiley's *AD Magazine*, Routledge's *Instabilities and Potentialities*, *Offramp*, *Acadia*, *TxA Emerging Design & Technology Journal*, *JAE*, *PLAT*, *Architect's Newspaper*, and Archinect.

Hans Tursack is a designer from Philadelphia. He received a BFA in studio art from the Cooper Union School of Art, and an M.Arch from the Princeton University School of Architecture. His writing and scholarly work have appeared in *Perspecta Journal*, *Pidgin Magazine*, *Thresholds Journal*, *Log*, *Dimensions*, Archinect, and the *Architects Newspaper*. He recently received the Pietro Belluschi Fellowship at the MIT School of Architecture + Planning, the Willard A. Oberdick Fellowship from the University of Michigan's Taubman College of Architecture, a MacDowell Colony Research Fellowship, and UDRF from Exhibit Columbus.

10 Finished pavilion on site, aerial view

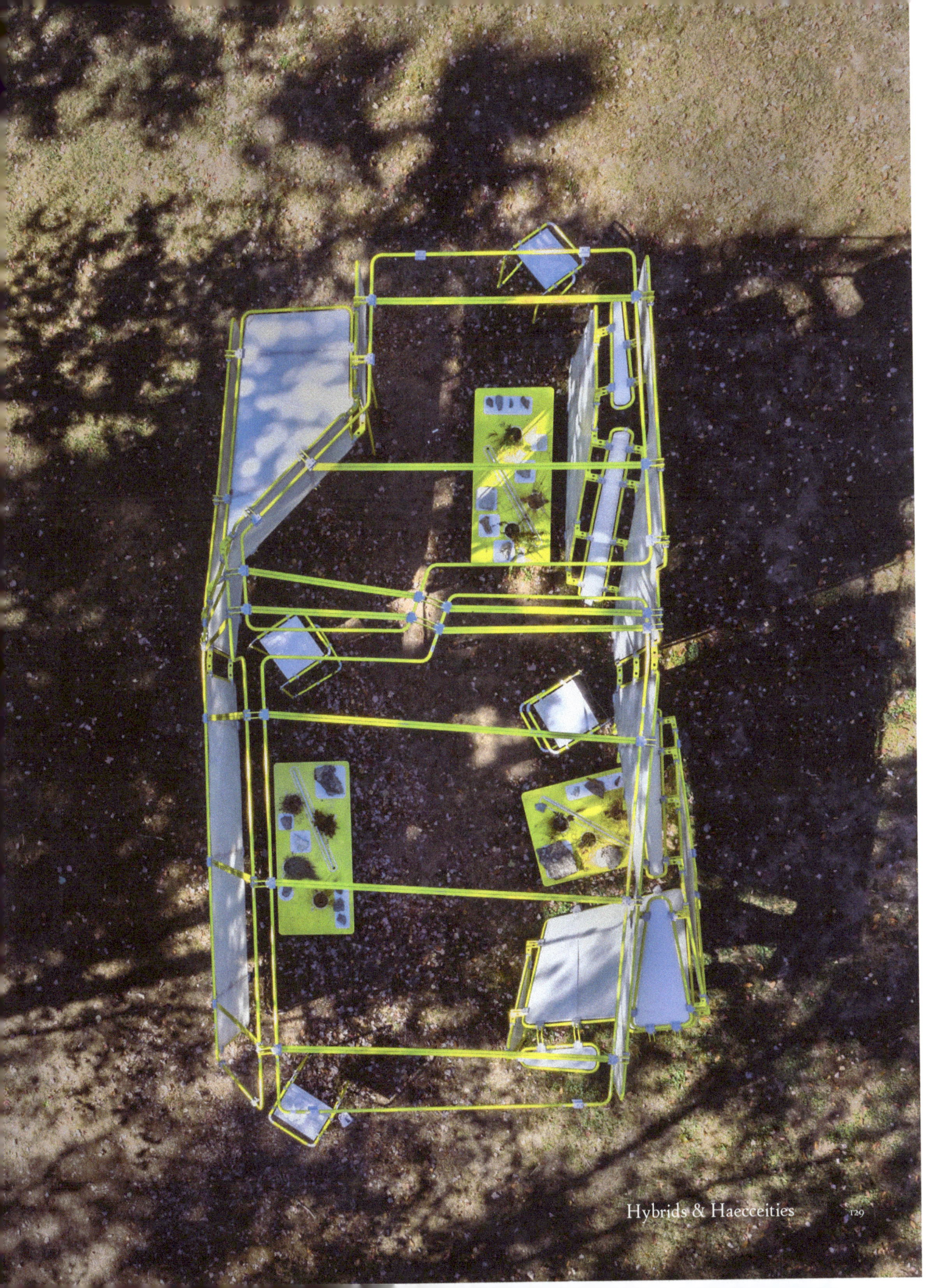

Shingle Nest

Fabricating Self-Supporting Shingle Envelopes Using Upcycled Elements

Tim Cousin

Shingle Nest is a project that addresses material upcycling from an architectural perspective. It aims to borrow materials from a recycling stream without compromising future cycles, effectively using architecture as a temporary keeper.

Significant challenges exist in achieving scalable processes for upcycling materials, including irregularity and deterioration, large tolerances, and difficulty in adapting these into precise and systematic digital fabrication workflows.

Shingle Nest addresses these challenges by leveraging the high tolerance of traditional shingling to accommodate non-standard upcycled plate elements in making architectural and structural surfaces while developing a material rejuvenation process that converts irregularity into aesthetic value. The proposed method eliminates the conventional need for a substructure and demonstrates the possibilities for precision and self-structuring assembly leveraging computational workflows.

The reclaimed pallets are broken and considered waste, yet many planks that compose them are still in good shape. The project starts by disassembling the pallet frames, sorting the planks, and returning the few broken ones to recycling streams. The salvaged ones are rejuvenated using traditional woodworking tools: a planer and a router are used to clean the boards' edges and one face. The second face of the boards is charred to retain thickness (with the planks warping, planing both sides would significantly reduce the material) and to register the engraving.

PRODUCTION NOTES

Architect: Tim Cousin

Status: Built

Site Area: 18 sq. ft.

Location: MIT Rotch Library

Date: 2021

1   Shingle Nest (Tim Cousin, 2021).

Hybrids & Haecceities

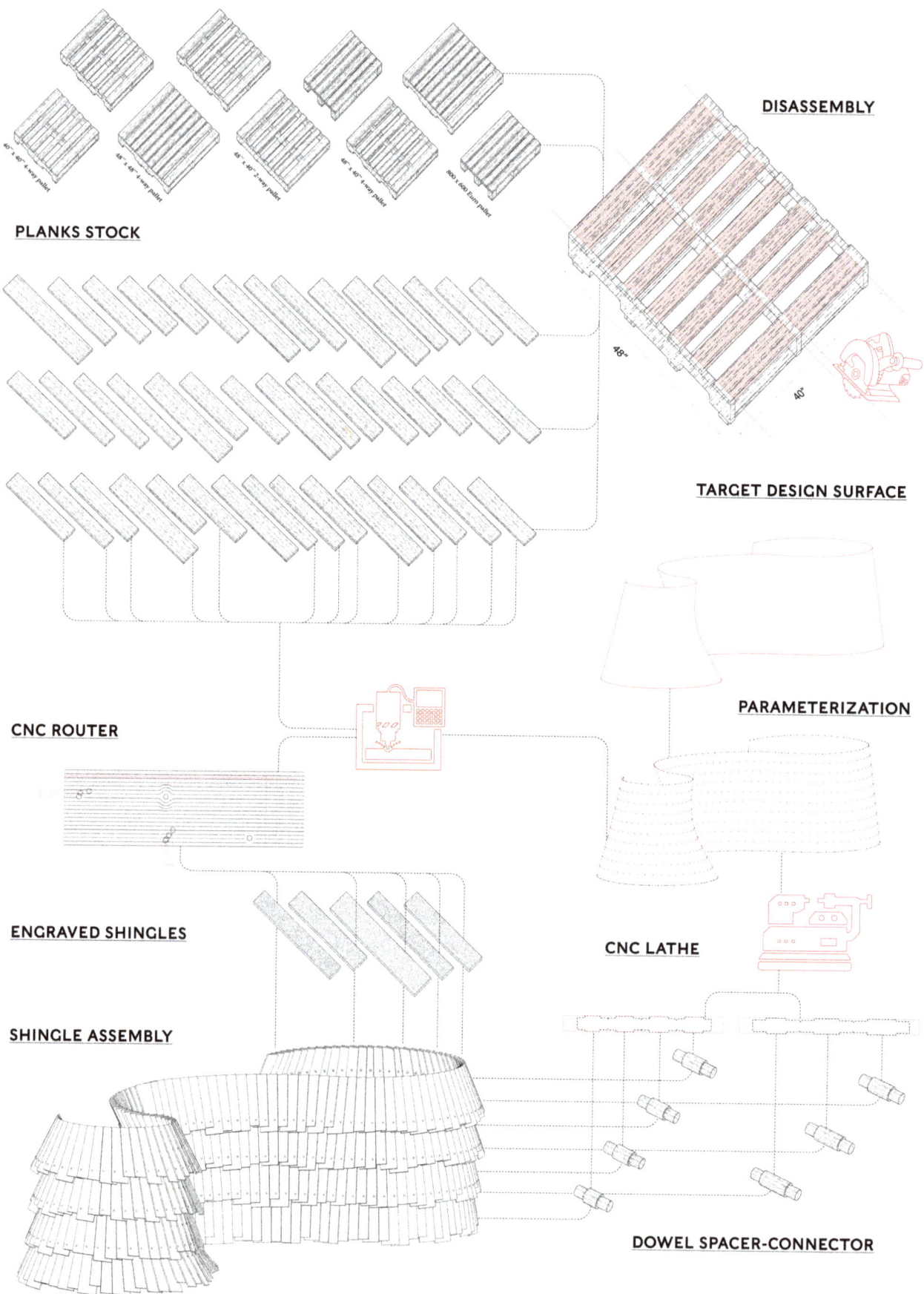

DISASSEMBLY

PLANKS STOCK

TARGET DESIGN SURFACE

PARAMETERIZATION

CNC ROUTER

CNC LATHE

ENGRAVED SHINGLES

SHINGLE ASSEMBLY

DOWEL SPACER-CONNECTOR

2 Overview diagram of the computational fabrication workflow

3   Salvaged pallets from the material recycling streams of MIT Campus

4   Planks reclaimed from the disassembly, before planing, routing and oiling

5   Cutting the dowel connectors on a CNC lathe

6   Engraving the planks on a CNC router

The assembly is a self-supporting shingle envelope that uses each plank as a structural tile punctually connected to its neighbors. This shingled envelope system is parametrized on a target surface in Rhino and Grasshopper that can be manipulated to produce variations of doubly-curved surfaces. The parametric model produces the shingle assembly using the stock information and outputs manufacturing instructions for the digital fabrication process.

The manufacturing process follows these steps:

- Wooden dowel connectors are cut using a CNC lathe; their shoulder lengths is directly outputted from the parametric model; it varies depending on the curvature of the envelope (and the subsequent inter-plank gap) throughout the shingles courses.
- Each plank is CNC-milled to locate the position of the connection points with the dowel connectors, which varies according to the envelope's curvature at each shingle location. In addition, a pattern is engraved onto

each plank with the CNC; it is consistent but registers very differently on each plank, revealing their singular nature: subtle unevenness results from their life of service as heavy-duty pallets.

- Holes for the connecting dowels are drilled at the locations engraved by the CNC. Due to the curvature of the target surface, these connections occur at an angle; these are outputted from the parametric model and drilled using a variable angle jig. Note that a CNC with more than three axes would allow merging this extra step into the previous step.
- The planks are assembled course by course from the bottom; all the positioning registration is embedded in the manufacturing of the pieces, both the plank holes and the shoulders of the dowel connectors.

In this first prototype, the shingle envelope wraps a small circular space, creating a nest. The inside of this shows the color variety of the different types of woods, oak, pine,

7 The charred, textured exterior and warm, rejuvenated wood interior of the shingle assembly

and chestnut, all finished with natural oil, creating a warm interiority. On the outer side, in stark contrast, the pavilion shows a rougher charred skin, animated by the singular carved patterns of the planks, that blends with the shingling rhythm in a rich textural effect.

The project advocates for a material ethic in design that values the essence of matter beyond its commodification and mobilize computational strategies, digital fabrication tools, and traditional woodworking towards the re-valuation of what is considered waste into a new compelling aesthetic proposition.

ACKNOWLEDGMENTS

This project was developed at MIT in the studio "Making Gestures, The Externalities of Knowledge, Methods and Materials in Fabrication Research" taught by Lavender Tessmer and Diego Pinochet with guest lecturers Maya Hayuk and Joseph Choma, with teaching assistant Gil Schwimmer Sunshine. The author acknowledges the precious help an critical teachings of this teaching team, as well as the invaluable support and assistance from Zain Karsan, Clara Copiglia, Chris Dewart, and Sacha Moreau.

REFERENCES

Brütting, J., G. Senatore, and C. Fivet. 2019. "Form follows availability; Designing structures through reuse." In *Journal of the International Association for Shell and Spatial Structures* 60(4): 257–265.

Becker, M., A. Fromm, P. Mecke, et F. Keller. 2020 "Casting Bespoke Connectors for Structural Shingles." In Advances in Architectural Geometry 2020.

IMAGE CREDITS

Figure 1-9: ©Tim Cousin 2021

Tim Cousin is a Graduate Architecture student at MIT, where he investigates narratives of material resources in the built environment. He is working on computational upcycling strategies to design wood structures with the Digital Structures group at MIT. Originally from Annecy in the French Alps, Tim studied applied arts in France and received a BSc.Arch from the Swiss Federal Institute of Technology (EPFL) in Lausanne, Switzerland. He spent the last year of his Bachelor's as a visiting student at the University of Tokyo. Before joining MIT, Tim collaborated with several architecture offices in Japan, Switzerland, Denmark, and Australia.

8 The Shingle Nest installed in the Rotch Library at MIT.

The CantiBox

Robotic Assembly of Interweaving Timber Linear Elements Using Bespoke Interlocking Timber-to-Timber Connections

Davide Tanadini, Giulia Boller, Pok Yin Victor Leung, Yijiang Huang, Pierluigi D'Acunto

In recent years, the building industry has increasingly promoted digital techniques, such as robotic manufacturing. Much scientific research has been conducted in the field of robotic assembly of timber structures, made of plates and linear elements (Robeller et al. 2017; Thoma et al. 2018; Leung et al. 2021; Hua et al. 2022). In this context, the robotically fabricated *CantiBox* project (Figure 1) presented here constitutes a novel application of the design and automatic assembly of interlocking timber-to-timber connections. The structure is composed of 60 linear elements of solid spruce interconnected through half-lap joints to form a reciprocal network.

GLOBAL DESIGN

CantiBox consists of three independent units: two lateral boxes connected to the ground and a cantilevering central box, which is simply supported by the other units. Each unit is composed of 20 timber linear elements with a cross-section of 10 by 10 cm, which are automatically assembled via robotic fabrication. Each face of the boxes is made of six timber linear elements that are arranged in the plane to generate a reciprocal system (Figure 2). Equilibrium-based methods for structural design and Finite Element Analysis (FEA) are used to evaluate the global structural behavior of the *CantiBox*. In the analysis, each joint is regarded as a rigid connection. Since the project is built outdoors, both the self-weight of the structure and the wind load are considered according to the Swiss norms (SIA 2021). For each connection, the most disadvantageous load combination is evaluated.

PRODUCTION NOTES

Project Coordination:	Davide Tanadini, Victor Pok Yin Leung, Yijiang Huang
Structural Design:	Davide Tanadini, Giulia Boller, Pierluigi D'Acunto
Mechatronics and Robotic Execution:	Victor Pok Yin Leung
Robotic Task and Motion Planning:	Yijiang Huang
Robotic Screwdriver Design:	Marco Rossi
Location:	Zurich, Switzerland
Date:	2022

1　One of the timber units during construction, with the robotic arm positioning a timber element following the assembly sequence (©Gramazio Kohler Research 2022).

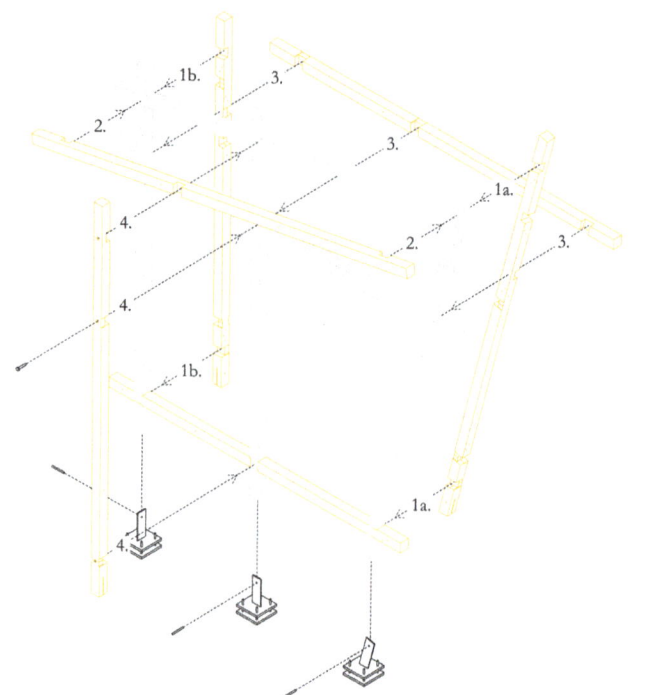

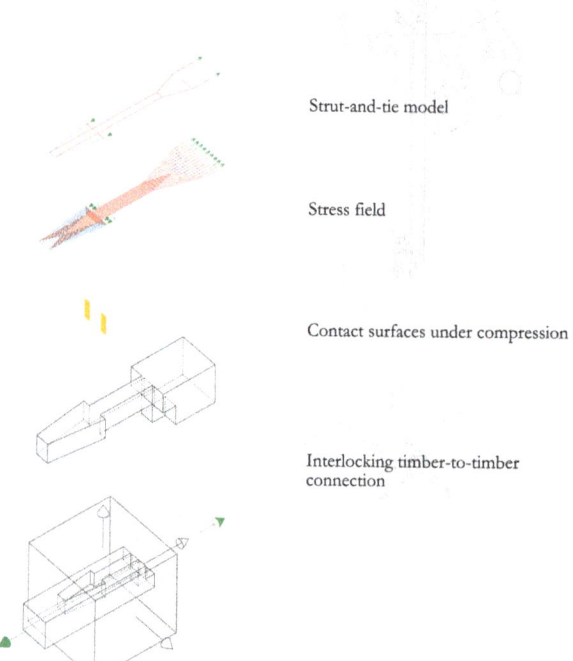

Strut-and-tie model

Stress field

Contact surfaces under compression

Interlocking timber-to-timber connection

2 The interweaving logic used during the assembly allows the number of metal fasteners to be limited to one per face. In fact, only the last assembled element, the key element, requires the use of a screw to close the reciprocal system and prevent the system from disassembly. This screw does not have any structural function.

3 Step-by-step structural analysis and design of a generic interlocking timber-to-timber connection. Based on the initial geometry, the contact surfaces under compression required for force transfer are identified. The flow of internal forces is represented by stress fields or equivalent strut-and-tie models. The geometry of the joint and stress field is adjusted such that stresses do not exceed the yield conditions.

CONNECTION DESIGN

The static method of limit analysis, based on plastic theory, is adopted to determine the capacity of the interlocking timber-to-timber connections and for their design (Tanadini and Schwartz 2021). In each of the lap joints, forces are transferred between the elements by means of compressed contact areas. The capacity depends on two parameters: the size of the contact surface and the strength value associated with it (Figure 3). The contact surface strength indicates the maximum stress that a surface can withstand. The strength value is calculated thanks to stress fields and timber yield conditions, and then validated through mechanical tests.

As each connection is subjected to different loads, it is possible to modify the contact surfaces required for load transfer by varying the geometry of the joints. The prototypical half-lap joint connection is therefore customized, whenever necessary, to adapt its capacity to the internal stresses. The parametric space for customization is carefully designed such that joints can be assembled by our robotic tools and are machinable by commonly available automatic joinery machines (Figure 7).

ROBOTIC ASSEMBLY

The structure is constructed through a fully automatic process (Leung et al. 2021), which uses a set of distributed robotic clamps (Figure 4) and screwdrivers (Figure 5) to operate in collaboration with an industrial robotic arm. Each of the three units is constructed spatially in an automatic process (Figure 1), as opposed to planar sub-assemblies. Robotic clamps are used for most joints that do not require fasteners. Robotic screwdrivers, which can be loaded with a left-in fastener, are used for the four key elements.

CONCLUSIONS

The *CantiBox* project demonstrates two cutting-edge approaches for jointed timber structures. First, the static method of limit analysis for timber-to-timber connection design, allows the adjustment of individual lap joint geometries based on real-time performance assessment. Second, distributed robotic tools are used to achieve a fully automatic assembly process. Both technologies complete an important knowledge gap that enables bespoke design and construction of spatial timber structures. Their flexibility

The CantiBox Tanadini, Boller, Leung, Huang, D'Acunto

4 The robotic arm positions two robotic clamps to insert a new timber element in the structure (©Gramazio Kohler Research 2022)

5 A key timber element is assembled by a robotic screwdriver loaded with a left-in fastener (©Gramazio Kohler Research 2022)

6 General view of the *CantiBox*; the structure is partially covered with translucent fabric to provide shading (©Lukas Ingold 2022)

7 All the bespoke connections are machined with a Hundegger Robot Drive automatic joinery machine (©Gramazio Kohler Research 2022)

8 Thanks to its inherent spatial reciprocal configuration, the *CantiBox* project achieves high geometric complexity using only simple planar connections (©Gramazio Kohler Research 2022)

to accommodate custom design can be witnessed in the use of the interweaving logic to create a reciprocal network (Figures 7, 8).

ACKNOWLEDGMENTS

CantiBox is the result of a research collaboration between Gramazio Kohler Research (ETH Zurich), the Chair of Structural Design (ETH Zurich), the Digital Structures Research Group (MIT), the Professorship of Structural Design (TU Munich), and the Institute for Lab Automation and Mechatronics (ILT, OST Rapperswil).

Special thanks for their significant support in the project go to: Prof. Fabio Gramazio (Gramazio Kohler Research, ETH Zurich), Prof. Matthias Kohler (Gramazio Kohler Research, ETH Zurich), Prof. Dr. Joseph Schwartz (Chair of Structural Design, ETH Zurich), Prof. Dr. Caitlin Mueller (Digital Structures, MIT Boston), Prof. Dr. Agathe Koller-Hodac (ILT, OST Rapperswil), Dr. Aleksandra Anna Apolinarska (Gramazio Kohler Research, ETH Zurich), Dr. Lauren Vasey (Gramazio Kohler Research, ETH Zurich), Gonzalo Casas (Gramazio Kohler Research, ETH Zurich), Philippe Fleischmann (NCCR Digital Fabrication, ETH Zurich), Michael Lyrenmann (NCCR Digital Fabrication, ETH Zurich), Leandro Nahuel Barroso (Chair of Structural Design, ETH Zurich), Rodrigo Mendoza Diaz (Chair of Structural Design, ETH Zurich), Dario Quaglia (Chair of Structural Design, ETH Zurich), Valentin Ribi (Chair of Structural Design, ETH Zurich), Louis Strologo (Chair of Structural Design, ETH Zurich), and Luca Steiner (ILT, OST Rapperswil).

REFERENCES

Hua, Chai, Zhixian Guo, Hans Jakob Wagner, Tim Stark, Achim Menges, and Philip F. Yuan. 2022. "In-Situ Robotic Fabrication of Spatial Glulam Structures." In *Proceedings of CAADRIA 2022*. Sydney: CAADRIA. 41–50.

Leung, Pok Yin Victor, Aleksandra Anna Apolinarska, Davide Tanadini, Fabio Gramazio, and Matthias Kohler. 2021. "Automatic assembly of jointed timber structure using distributed robotic clamps." In Proceedings of CAADRIA 2021. Hong Kong: CAADRIA. 583–592.

Robeller, Christopher, Yves Weinand, Volker Helm, Andreas Thoma, Fabio Gramazio and Matthias Kohler. 2017. "Robotic integral attachment." In *Fabricate 2017*. London: UCL Press.

Swiss Society of Engineers and Architects. 2021. *SIA 260: Basis of Structural Design; SIA 261: Actions on Structures*.

Tanadini, Davide and Joseph Schwartz. 2021. "Analysis and design of timber-to-timber connections based on the lower bound theorem of the theory of plasticity." In Proceedings of the WCTE 2021. Santiago: WCTE.

9　The *CantiBox* project leverages innovative approaches to structural design and robotic assembly to enable the development of a customized, spatial timber structure (©Lukas Ingold 2022)

Thoma, Andreas, Arash Adel, Matthias Helmreich, Thomas Wehrle, Fabio Gramazio, and Matthias Kohler. 2018. "Robotic Fabrication of Bespoke Timber Frame Modules." In *Robotic Fabrication in Architecture, Art and Design 2018*. Cham: Springer. 447–458.

IMAGE CREDITS
Figure 5, 7, 10-12: ©Gramzio Kohler Research 2022
Figure 4-8: ©Lukas Ingold 2022
All other drawings and images by the authors

Davide Tanadini is a lecturer and PhD student at the Chair of Structural Design at ETH Zurich. His research focuses on possible applications of the theory of plasticity and graphic statics on timber structures and timber joints, and their implementation on digital fabrication and robotic assembly. Davide graduated from ETH Zurich as a civil engineer in January 2018.

Giulia Boller is a scientific assistant and PhD student at the Chair of Structural Design at ETH Zurich. She is both an engineer and an architect. She gained professional experience at Renzo Piano Building Workshop (RPBW). Giulia graduated with honours in Building Engineering and Architecture at the University of Trento in 2015.

Pok Yin Victor Leung is a PhD student at the Gramazio Kohler Research group at ETH Zurich. He received his BA in Architectural Studies from HKU in 2011 and MSc in Architectural Studies (in Design and Computation) from MIT in 2016. His expertise includes timber joint design, structural design, software development, mechatronics, robot control, and motion planning.

Yijiang Huang is a recently graduated PhD student from the Building Technology program at the Massachusetts Institute of Technology, advised by Prof. Caitlin Mueller. His research focuses on using automated planning techniques to streamline the design-to-execution pipeline of robotic assembly of spatial structures.

Pierluigi D'Acunto graduated in Building Engineering-Architecture from the University of Pisa in 2007 and received a Master of Architecture from the AA School of Architecture in London in 2012. In 2018, he completed his PhD with distinction at ETH Zurich. Pierluigi is currently tenure-track Assistant Professor of Structural Design at the Department of Architecture of the Technical University of Munich (TUM).

The Arroyo Bridge

Collaborative Robotics for Large Scale Construction

R. Scott Mitchell

The Arroyo Bridge, spanning 80-feet across a ravine near Los Angeles, is anchored by four, 35-foot deep concrete pilings, hand-bored into the hillside to provide adequate support in the alluvial soil. An asymmetrical superstructure of over 600 discrete parts, the bridge is joined by unique part-to-part connections that frame the pedestrian walkway.

From a distance, the biomorphic shape integrates into a steep landscape of eucalyptus, coming into view only upon approach. The tubular silhouette draws inspiration from the vascular architecture of leaves and layered branching system of trees in the neighboring forest.

This project is the product of a 6-year research initiative. The pedestrian span was originally conceived as a speculative design project during a design-build studio at the USC School of Architecture in 2014. It was later realized using a pioneering computational design and manufacturing process with humans and robots working side-by-side.

PRODUCTION NOTES

Designer:	R. Scott Mitchell
Client:	Mary and David Martin
Status:	Built
Location:	Los Angeles
Date:	2020

1 Completed bridge span (Iwan Baan, 2021)

2   Bridge view from west (Iwan Baan, 2021)

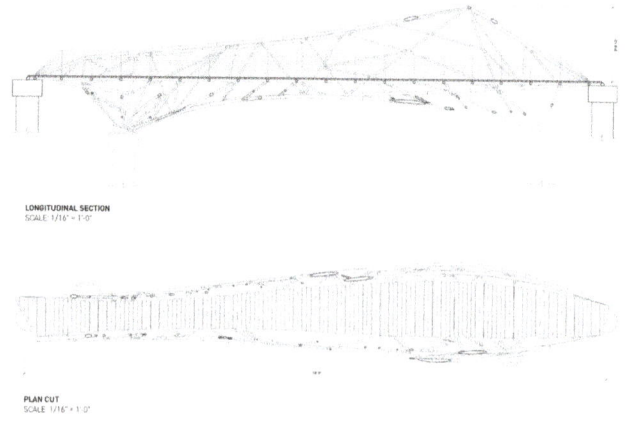

LONGITUDINAL SECTION
SCALE 1/16" = 1'-0"

PLAN CUT
SCALE 1/16" = 1'-0"

3   Longitudinal section and plan cut (Gigante AG, 2019)

4   Robotic positioning for pile cap fabrication, Los Angeles (Gigante AG, 2019)

While computational tools and digital fabrication processes informed the numerous iterations of the bridge, progressively enhancing both its structural integrity and architectural design, the bridge's complex geometry made it impossible and prohibitively expensive to build with traditional manual fixturing methods.

Due to limited site access, the design team developed an innovative plan to build the bridge as components. Automating the assembly, in which parts are held in programmed robotic arms, enabled the welders to enter the workspace, making the bridge a unique study in collaborative robotics.

Using readily available parametric tools, the team identified optimal component breakouts and divided the bridge into 30 unique nodes. Parameters for each node were based on fixture orientation, transport volumes and an assembly weight of approximately 500 lbs (227 kg). Working initially

with KUKAprc on the university's robots, the team prototyped fabrication workflows. Upon partnering with the Autodesk Robotics Lab (ARL), they developed a streamlined technique utilizing Mimic for Maya for simulation and robot pick-and-place code generation.

The design/fabrication team was awarded a residency at the Autodesk Technology Center in Boston. There they developed custom magnetic end-effectors and tooling for measurement and member placement. Using computational and fabrication tools, the team instructed robot arms to hold each part in position for human welders. The robot arm operated as a positioning and metrology device, eliminating formwork and material waste in order to minimize environmental impact. On average, it took the team thirty minutes per node to translate the geometric models into fully simulated and programmed instructions for the robot

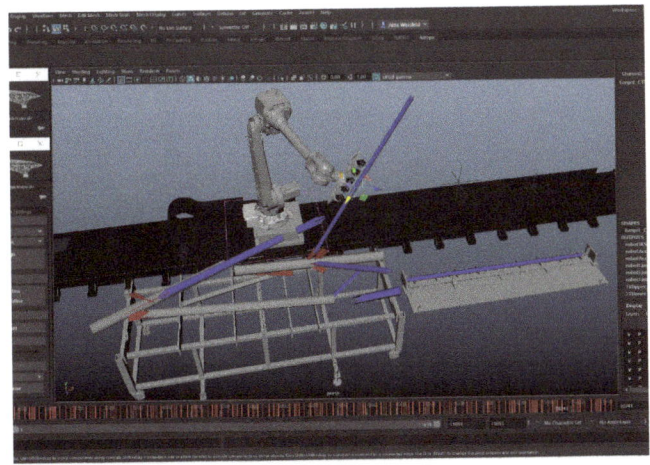

5   Robotic positioning simulation (Gigante AG, 2019)

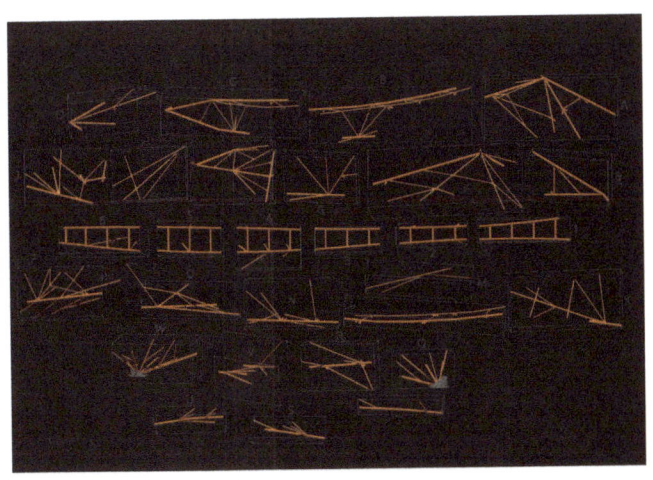

7   Bridge component nodes (Gigante AG, 2019)

6   Robotic positioning during fabrication, Boston (Gigante AG, 2019)

8   Node test assembly, Hawthorne (Gigante AG, 2020)

arm. All robotic prefabrication was achieved with under one millimeter of tolerance, far more precise than the specified allowable limit of six millimeters.

Once positioned, each steel member was tack welded and braced for shipment. Following local codes, all structural welding had to be completed in a Los Angeles city-approved facility. The bridge was test assembled for the first time at a welding yard in Hawthorne, California. Although no two parts had ever been mated, the nodes fit together perfectly due to the closely defined tolerances. The complexity of the joints required the development of new welding techniques. Abutment caps, anchor plates, decking, and railing components were fabricated simultaneously. Working with Leica Geosystems, a total station and custom prismatic tooling were used to verify the accuracy of all shop and site work.

Nodes were delivered to site as needed. An all-terrain tele-handler forklift transported the nodes up the steep ravine. A scaffolding system in the canyon created a construction platform, where each node was received, locked in place and verified. A proprietary connector design allowed the alignment of critical pieces, limiting on-site welding to less than ten percent of the total job. The entire base span was installed in one week, reducing expected assembly time by more than fifty percent.

ACKNOWLEDGMENTS

Design/Fabrication Team: Adan Macias, Alex Weisfeld, Diana Yan
Autodesk Robotics Lab: Evan Atherton, Heather Kerrick
Mary and David Martin
MADWORKSHOP
Autodesk Technology Center, Boston
USC School of Architecture

9   Bridge installation (Gigante AG, 2020)

10   Worm's eye view (Iwan Baan, 2021)

REFERENCES

Bechthold, Martin. 2010. "The Return of the Future: A Second Go at Robotic Construction." *Architectural Design* 80(4): 116–21.

Helm, Volker. 2014. "In-Situ Fabrication: Mobile Robotic Units on Construction Sites." *Architectural Design* 84(3): 100–07.

Kerber, Ethan, Tobias Heimig, Sven Stumm, Lukas Oster, Sigrid Brell-Cokcan, and Uwe Reisgen. 2018. "Towards Robotic Fabrication in Joining of Steel." In *Proceedings of the 35th International Symposium on Automation and Robotics in Construction and Mining (ISARC)*. Berlin, Germany.

Kolarevic, Branko, and José Pinto Duarte. 2018. *Mass Customization and Design Democratization*. Abingdon: Routledge.

Stumm, Sven, Johannes Braumann, and Sigrid Brell-Cokcan. 2016. "Human-Machine Interaction for Intuitive Programming of Assembly Tasks in Construction." *Procedia CIRP* 44: 269–74.

Thoma, Andreas, Arash Adel, Matthias Helmreich, Thomas Wehrle, Fabio Gramazio, and Matthias Kohler. 2019. "Robotic Fabrication of Bespoke Timber Frame Modules." In *Robotic Fabrication in Architecture, Art and Design (ROBARCH) 2018*, edited by J. Willmann et al. Cham: Springer.

11 Robotic weld positioning, Boston (Gigante AG, 2019)

IMAGE CREDITS

R. Scott Mitchell is the owner/principal of Gigante AG, a Los Angeles design-build and fabrication consulting firm. He has worked for Gehry Partners, Morphosis, Bestor Architecture, and Atelier Van Lieshout. Early in his career he worked as both a scientific instrument maker and a structural fabricator. From 2018 to 2020 he was a research resident at the Autodesk Technology Center in Boston. He is Associate Professor of Practice at the USC School of Architecture, where he has been teaching design and digital fabrication since 2007. He holds an MArch from SCI-Arc and a BA in Architectural Studies from Brown University.

Responsive Spatial Print Trajectory

3D Printing of Clay Lattices with Self-Corrective Recalibration

Francisco Jung, Sulaiman Al Othman, Hyeonji Claire Im,
Jose Luis García del Castillo y López, Martin Bechthold

Additive manufacturing of clay has gained interest in and ushered developments for the application of building architectural components due to the wide availability and various environmental benefits of clay. The most common printing process typically relies on a layer-by-layer deposition technique that is very slow and results in increased operational costs. In addition, current workflows do not take into account material-specific properties and lack the data required to anticipate non-linear deformations throughout the printing process. This poses a particular challenge for large-scale applications because the effects of material behavior amplify exponentially with scale. This project presents a novel method of spatially printing clay lattices by controlling fabrication parameters such as the printing head speed and the material extrusion rate following a 3D-choreographed toolpath. Spatial printing refers to the unrestricted movement of the printer nozzle in three axes (*x, y, z*) when extruding material, as opposed to the conventional 2-axis layer-by-layer deposition that is very slow and results in increased operational costs. This method—enhanced with an integrated industrial laser displacement sensor to collect deflection data subsequently used to calibrate the next layer toolpath geometry in real-time—works optimally with carbon-fiber reinforcements for increased tensile performance. This integrated workflow was utilized to produce a 2.1-meter self-supporting structure in a few days using less materials while preserving all intended architectural qualities.

PRODUCTION NOTES

Designer: MaP+S Group

Sponsor: Kuwait Foundation for the Advancement of Sciences

Status: Complete

Location: Harvard University

Date: 2019

1   2.1-meter high lattice prototype demonstrating the viability of the spatial print trajectory (SPT) 3D printing method for fiber-reinforced clay

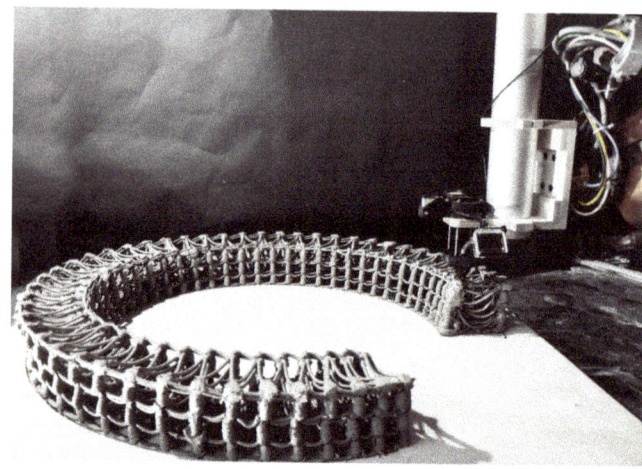

5

3

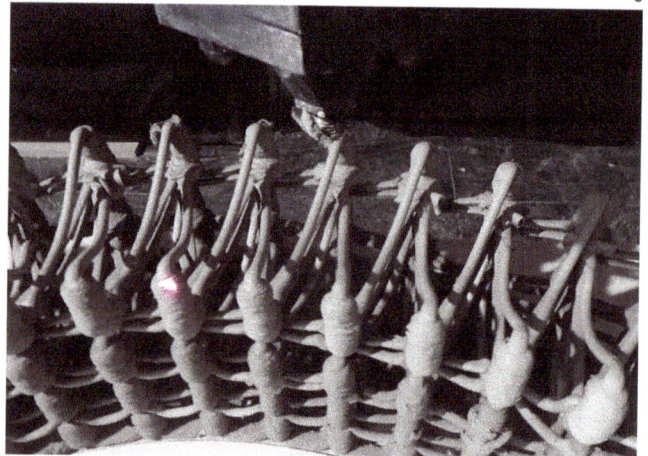

4

2 6-axis robot arm with customized clay extrusion system for continuous printing

3 The lattice vertical pylons are printed at a low head speed with high material feed rate

4 An industrial laser displacement sensor calibrates the next toolpath layer

5 The lattice horizontal ties are printed at a higher head speed with low material feed rate

SPATIAL PRINT TRAJECTORY

Two different printpath patterns alternate to simulate actions that complement material behavior: anchoring and dragging to produce a lattice unit in the air without support (Al Othman et al. 2018). The purpose of the first printpath pattern is to build vertical pylons of the lattice with less material, requiring the printer to deposit material at a low printing speed with high material extrusion rate in the vertical direction (Figure 3). Then, the nozzle drags the material at a higher speed to the adjacent side of the trajectory to tie the previously built pylon with consecutive pylons like a space truss, making one lattice unit. The lattice unit is then superimposed by the second printpath pattern to create a horizontal layer bracing the pylons with a foundation and horizontal ties (Figure 5). Several units can be repeated and controlled dimensionally to create a spatial print trajectory, which can be stacked to produce complex forms with different characteristics (Figure 8).

RESPONSIVE COMPONENT: SELFCORRECTIVE CALIBRATION

Major deflections, elevated degrees of distortion, and retracting occurred during deposition due to the swelling potential of clay compounded by the mechanically induced actions of dragging and drying. Incorporation of an industrial laser displacement sensor was necessary to measure the deflections and distortions in order to correct the digital geometry and calibrate the z-coordinate of the consecutive lattice toolpath's first point in real time (Figure 4). This closes the loop of the feedback system between the physical state of the material and its digital geometry, permitting the deposition process to readjust in real-time for any discrepancy without human intervention (Claire Im et al. 2019).

Responsive Spatial Print Trajectory Jung, Al Othman, Im, García del Castillo y López, Bechthold

6 RSPT incorporates geometric parameter control to design different architectural qualities

MATERIAL ENHANCEMENT

The combination of natural and forced drying amplified the problem of cracking. Therefore, several strategies were researched to enhance the wet material properties and improve the mechanical performances of the clay lattice. As a result of extensive trials, the addition of carbon fiber to the clay mix increased its structural capacity in bending and tension. It is worth noting that the carbon fiber reinforcements seemed to mitigate deformation or warping due to drastic temperature and humidity differential induced by forced heating.

2.1-METER HIGH SEMI-ENCLOSURE

The final lattice structure with a dimension of 1.2 x 1.2 x 2.1 meters was printed using a 6-axis industrial robot arm in less than five days, demonstrating the capacity of the updated methodology to successfully scale up geometries while conserving time and material. The height of each lattice level gradually increases with the level or layer count to decrease the material density on upper layers, allowing more light and air to filter in while reducing loads (Figure 8).

ACKNOWLEDGMENTS

This research was funded by the Kuwait Foundation for the Advancement of Sciences (KFAS) under project code "CB18- 65EA-01." The authors would like to sincerely thank Harvard University Ceramics for the valuable support.

REFERENCES

Al Othman, Sulaiman, Hyeonji Claire Im, Francisco Jung, and Martin Bechthold. 2019. "Spatial Print Trajectory." In *Robotic Fabrication in Architecture, Art and Design 2018*, edited by J. Willmann, P. Block, M. Hutter, K. Byrne, and T. Schork, 167–80. Cham: Springer International Publishing. https://doi.org/10.1007/978-3-319-92294-2_13.

Im, Hyeonji Claire, Sulaiman Al Othman, and Jose Luis García del Castillo. 2018. "Responsive Spatial Print; Clay 3D Printing of Spatial Lattices Using Real-Time Model Recalibration." In *ACADIA '18: Recalibration, On Imprecision and Infidelity; Proceedings of the 38th Annual Conference of the Association for Computer Aided Design in Architecture*. ACADIA: Mexico City, Mexico. 286–293. https://doi.org/10.52842/conf.acadia.2018.286.

IMAGE CREDITS

Figure 1 - 9: © MaP+S Group

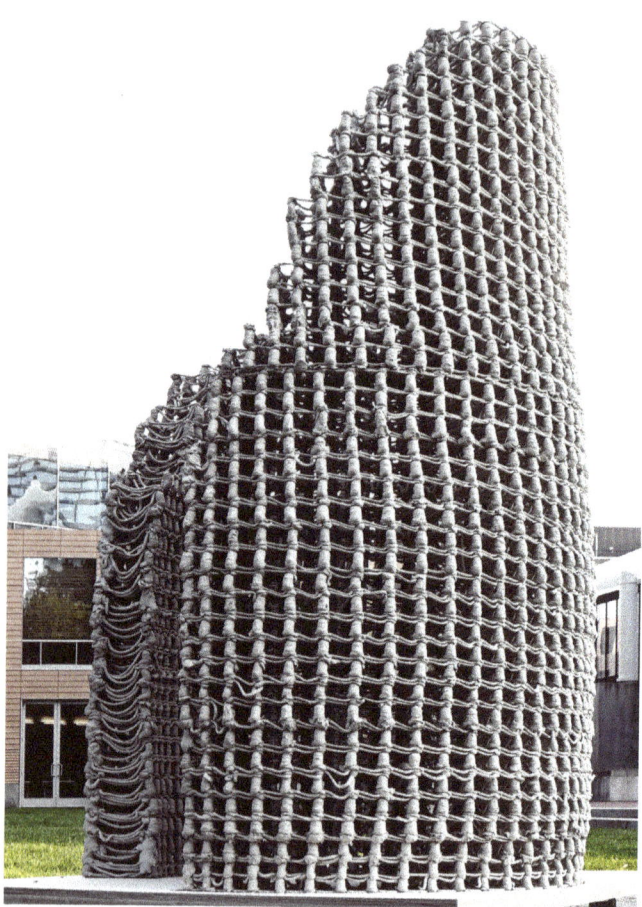

6   Full-scale lattice semi-enclosure

7   Full-scale lattice semi-enclosure.

Francisco Jung is a researcher with a background in architecture, robotic fabrication, and aerospace engineering. He received a BArch at Syracuse University, MSAUD at Columbia University, and MDes Technology at Harvard University. He practiced as an urban designer at the New York City Public Design Commission and the American Planning Association. He received certification from MIT's Computer Science and Artificial Intelligence Lab. He also worked at NASA Johnson Space Center where he designed and prototyped spacecraft for the Artemis Program (ex-Cislunar Gateway).

Sulaiman Al Othman is a Doctor of Design candidate at the Harvard Graduate School of Design. His research focuses on developing predictive models to address material uncertainty in additive manufacturing of earth-based materials for various design applications.

Hyeonji Claire Im Throughout her career over 10 years as an architect, she has been involved in various projects on a wide range of scales, locations, and functions. Her experience made her believe in the need for innovative construction methods using robotic fabrication and computational workflow that improve the built environment. She was a researcher at Harvard University, MaP+S Group with a master's degree in Design Studies at Harvard University Graduate School of Design. She is currently a senior project architect at Perkins&Will.

José Luis García del Castillo y López is an architect, computational designer, and educator. His work focuses on the development of digital frameworks that help democratize access to creative technology for designers and artists. Jose Luis is a registered architect and holds a Doctorate in Design and a Master in Design Studies on Technology from the Harvard University Graduate School of Design, where he is currently Lecturer in Architectural Technology at the Material Processes and Systems Group (MaP+S). He also leads ParametricCamp, an online platform for open knowledge in computational design.

Martin Bechthold is the Kumagai Professor of Architectural Technology at the Harvard Graduate School of Design. He co-directs the Master in Design Engineering program at Harvard and is the founding director of the Materials Processes and Systems (MaP+S) Group. MaP+S advances the understanding of materiality in the built environment, leveraging design computation, robotics, and pursuing collaborative work with material scientists. Bechthold also founded Harvard's Laboratory for Design Technologies as a platform for industry and academy to connect in the effort to expand knowledge in design and construction. His work has been widely published nationally and internationally.

8   An additional horizontal layer reduces the stresses caused by self and forced-heat drying, which led to cracks

Responsive Spatial Print Trajectory Jung, Al Othman, Im, García del Castillo y López, Bechthold

UNLOG

A Deployable and Lightweight Timber Frame

Leslie Lok, Sasa Zivkovic

Easily deployed and assembled, *UNLOG* unfolds several logs into an undulating and lightweight timber A-frame structure through robotic kerfing and bending-active kinematics (Schleicher et al. 2014). The installation (Figures 1, 2) provokes new methods of framing for timber construction. The project consists of three A-frame modules measuring 8ft x 8ft, which are connected by radially splayed members. The three modules orient themselves towards three nearby large trees, generating a massing with varied spatial qualities (Figure 5). The structure incudes two benches for seating and is cladded by 1/8 inch ash sheathing members. The sheathing panel, in addition to the kerfed wall component, resists lateral forces on the structure.

RE-FRAMING MATERIAL USAGE

Today over 90% of all new single-family homes in the United States are constructed with transported soft-wood species (Dietz 2018). *UNLOG* questions current paradigms of logging by working with local forest resources and hard-wood trees devastated by the ongoing Emerald Ash Borer (EAB) epidemic (Flower et al. 2013). Expanding upon prior ash wood fabrication research (Zivkovic and Lok, 2020), *UNLOG* utilizes robotic kerfing techniques and Mixed Reality (MR) to transform dying ash trees into a materially efficient, valuable resource. Using only 6 1/2 logs, threaded rods, and custom recycled HPDE washers (Figure 3), *UNLOG* investigates how far a single log can be stretched—both literally as an assembly and figuratively as a resource.

KERFING METHOD

Kerfing is a relief cut technique that induces flexibility into

PRODUCTION NOTES

Designer:	HANNAH
Exhibition:	Biomaterial Building Exposition
Status:	Completed
Site Area:	355 sq. ft.
Location:	Charlottesville, VA
Date:	2022

1   Top view of *UNLOG*, a spatial pavilion

2   Aerial view demonstrates the integration of public seating within A-Frame modules

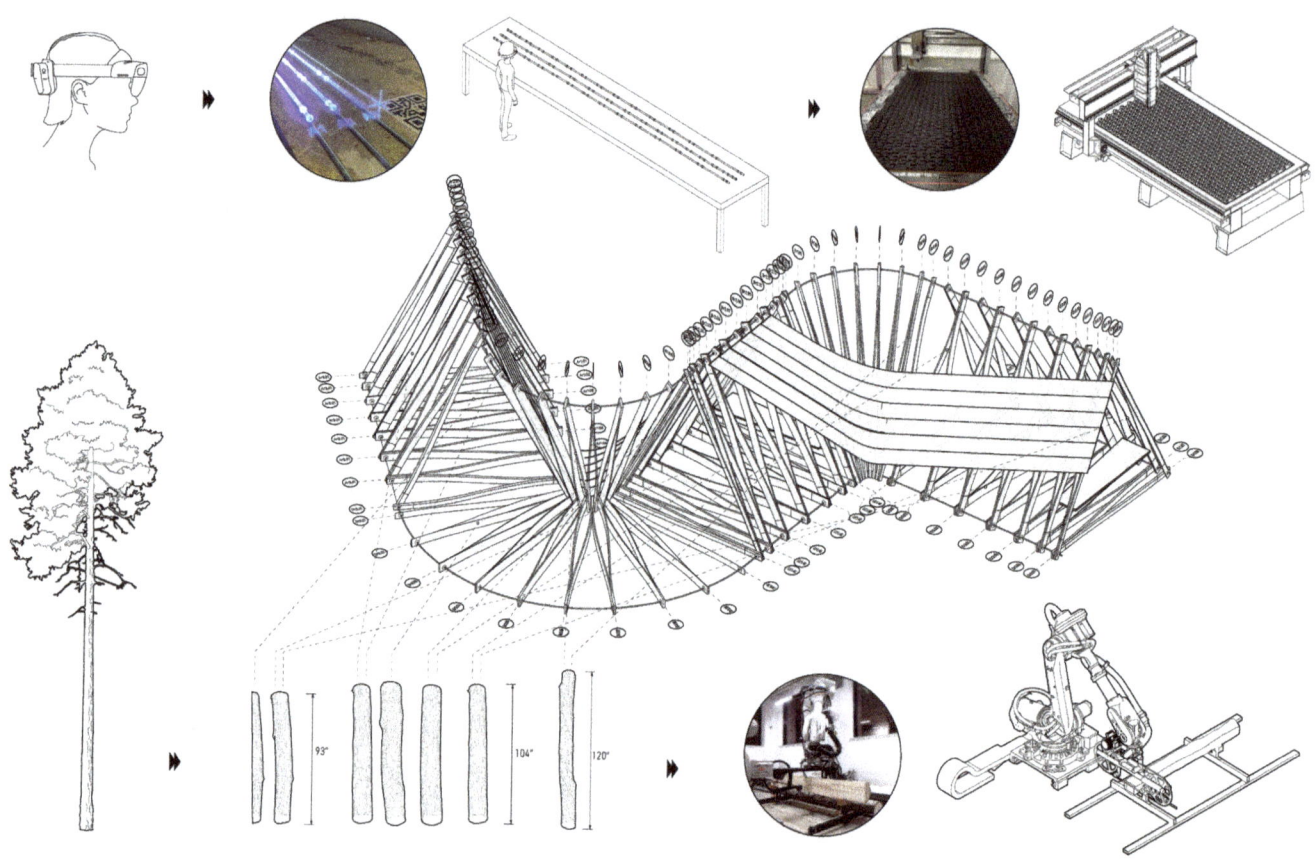

3 *UNLOG* workflow diagram showing robotically kerfed logs, while hex nuts are placed on threaded rod with custom-made recycled HDPE slip washers

a material. Though kerfing is often applied to a sheet material (Menges 2011), *UNLOG* employs methods to robotically kerf a log along its length, subsequently, unraveling the volumetric roundwood into an operable leaf-spring component. Like the Torus Research Pavilion by CODA (Tornabell et al. 2014), the stretched kerfed log component has inherent bending-active structural principles (Schleicher et al. 2014). Geometric relationships, material constraints, board thicknesses, and log geometries for *UNLOG* were developed through physical prototypes, computational models, and drawing.

FABRICATION PROCESS

A 5 hp bandsaw end effector on a ABB IRB 6700 with a 6200 mm external track was used to process the logs with diameters between 11 and 14 inches. First, the logs were cut in half along the longitudinal axis with the robotic arm, then each half was rotated 90° and the ends of the log were squared off. Finally, an alternating pattern of 3/4 inch thickness, with

a 12 inch offset at either end was cut (Figure 4). The kerfing patterns for *UNLOG* were regular, but the robotic method enables a variety of cut profiles with varied thickness and curvature, which can be employed to structurally enhance the kerfed log component (Figure 5).

DETAILING & AR ASSEMBLY

The kerfed logs were pre-drilled with a 1-1/4 inch hole and stretched along 1/2 inch diameter threaded rods with 3/4 inch hex nuts that were pre-located with Mixed Reality (MR) instruction (Figure 6). The use of MR enables the fabricator to precisely place each hex nut at variable locations along the rod without templates or other measuring devices. The kerfed logs were fixed into location along the rod with a custom recycled HDPE slip washer (Figures 8, 9). Augmented Reality (AR) protocols adapted from prior work (Lok et al. 2021) were used to adjust and calibrate the spacing of nuts on the threaded rods and inspect the accuracy of the overall physical assembly against its digital parallel.

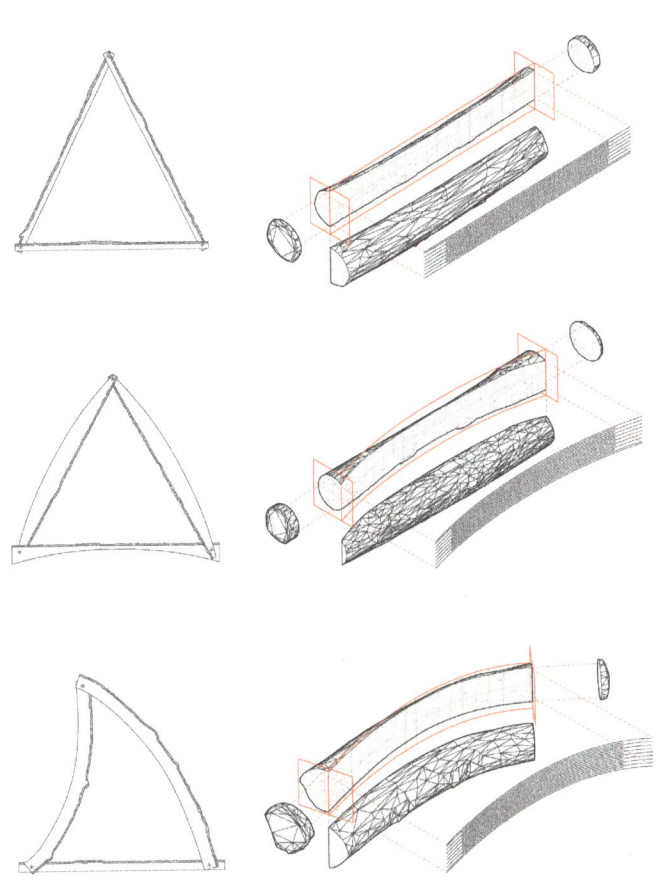

4   6-axis robot with bandsaw end-effector processing alternating kerf cuts

Mixed Reality Headset
QR Code
Coordination Point
Digital Hex Nut Location
Threaded Rod ID

5   Robotic cutting diagram for uniform profile, custom profile, and irregular profile

6   MR instruction is used to prefabricate the assembly of hex nuts on threaded rods

7   Eye level view illustrates the parametric spacing of timbers and the interior/exterior relationship of the sheathing panels

Hybrids & Haecceities

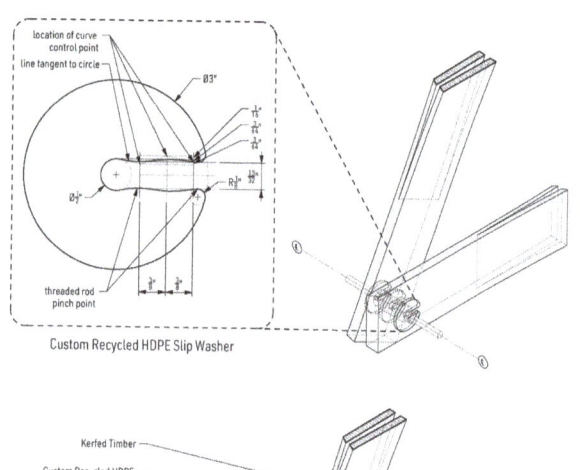

9 Recycled HDPE slip washer snaps onto threaded rod to fix timber in location

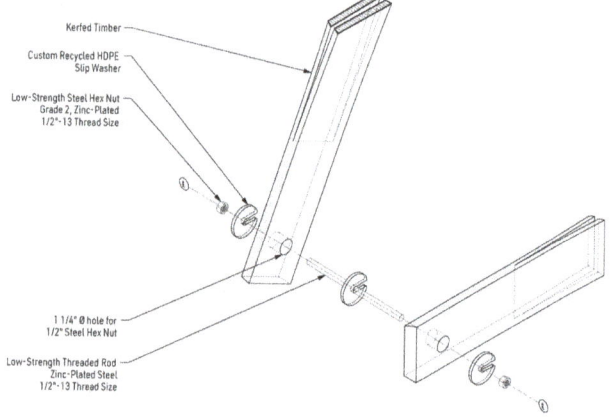

Custom Recycled HDPE Slip Washer

Kerfed Timber

Custom Recycled HDPE Slip Washer

Low-Strength Steel Hex Nut Grade 2, Zinc-Plated 1/2"-13 Thread Size

1 1/4" Ø hole for 1/2" Steel Hex Nut

Low-Strength Threaded Rod Zinc-Plated Steel 1/2"-13 Thread Size

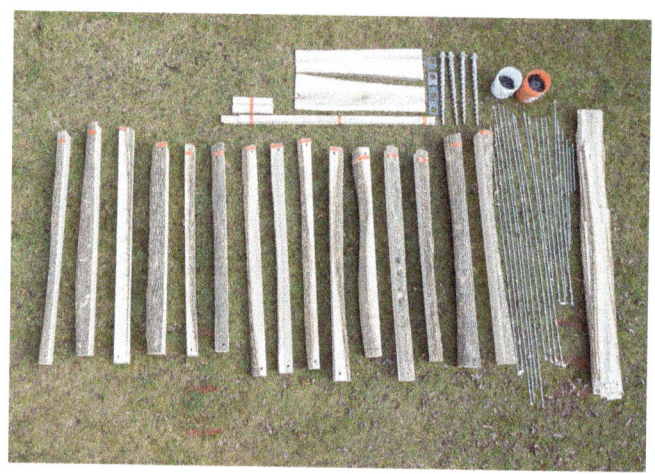

8 Assembly detail illustrating the assembly sequence and the custom slip-on washers

10 Assembly components

With a total weight of less than 2000 lbs, including hardware, the individual components were shipped in a flat-packed state (Figure 10) and unfolded at the installation site. From *UNLOG* to log, the pavilion can be easily disassembled and re-installed, thus presenting an easily transportable, deployable, and minimal-waste construction method.

ACKNOWLEDGMENTS

UNLOG is a commissioned pavilion for the Biomaterial Building Exposition at University of Virginia curated by Katie MacDonald and Kyle Schumann. HANNAH project leadership: Leslie Lok, Sasa Zivkovic. Project Manager: Lawson Spencer. UVA Project Manager: Collette Block. Team: Sahil Adnan, Shuo Feng, Lauren Franco, Alexander Kyaw, Shujie Liu, Yehong Mi, Shenkung Yang, Chi Zhang, Andreya Zvonar. Special thanks to the UVA Fabrication Lab Facilities team, workshop participants, and student volunteers for the assembly and disassembly of the pavilion. The Full Log Kerfing method was first developed by Savannah Chasing Hawk in the Timber Villa Option Studio at Cornell University in 2017. Project realized with scientific support from the Cornell Robotic Construction Laboratory (RCL) and the Cornell Rural-Urban Building Innovation Lab (RUBI Lab). This project received funding support from the Jefferson Trust, UVA Center for Global Inquiry & Innovation, UVA School of Architecture, and Cornell AAP Dean's Professional Development Fund.

REFERENCES

Dietz, Robert. 2019. "Framing Methods for Single-Family Homes, 2018: Eye On Housing." *Eye On Housing: National Association of Home Builders Discusses Economics and Housing Policy*, October 10, 2019. https://eyeonhousing.org/2019/10/framing-methods-for-single-family-homes-2018/.

Flower, C.E., K.S. Knight, and M.A. Gonzalez-Meler. 2013. "Impacts of the emerald ash borer (*Agrilus planipennis Fairmaire*) induced ash (*Fraxinus spp.*) mortality on forest carbon cycling and successional dynamics in the eastern United States." *Biological Invasions* 15 (4): 931–944.

Lok, Leslie, A. Samaniago, and L. Spencer. 2021 "Timber De-standardized: A Mixed Reality Framework for the Assembly of Irregular Tree Log Structures." In *ACADIA 21: Realignments toward Critical Computation; Proceedings of the 41st Annual Conference of*

12 Connection at radial components

13 Robotically sliced 3mm thick ash wood sheathing panels

11 View showing both material edges, the precise profile cut side of the log and the natural exterior with bark

the Association for Computer Aided Design in Architecture, edited by K. Dorfler, S. Parascho, and J. Scott. Online: ACADIA.

Menges, Achim. 2011. "Integrative Design Computation: Integrating Material Behaviour and Robotic Manufacturing Processes in Computational Design for Performative Wood Constructions." In *ACADIA 11: Integration through Computation; Proceedings of the 31st Annual Conference of the Association for Computer Aided Design in Architecture*. Banff, Alberta: ACADIA. 72–81

Schleicher, Simon, J. Lienhard, S. Poppinga, T. Masselter, T. Speck, and J. Knippers. 2011. "Adaptive Façade Shading Systems Inspired by Natural Elastic Kinematics." In *Proceedings of the International Adaptive Architecture Conference (IAAC)*. London: IAAC. 2545–54.

Tornabell, P., E. Soriano, and R. Sastre. 2014. "Pliable structures with rigid couplings for parallel leaf-springs: a pliable timber torus pavilion." *Mobile and Rapidly Assembled Structures IV* 136 (117): 485.

Zivkovic, Sasa, and Leslie Lok. 2020 "Making Form Work - Experiments along the Grain of Concrete and Timber" In *FABRICATE 2020: Making Resilient Architecture*, edited by J. Burry, J. Sabin, B. Sheil. London: UCL Press. 116–123.

IMAGE CREDITS

Figure 1, 2, 7, 9, 11: © Nana Iso, 2022
All other drawings and images by the authors

Leslie Lok is a co-founder at HANNAH and Assistant Professor at Cornell University, Department of Architecture. Her experimental design practice, HANNAH utilizes innovative forms of construction to advance building practices. At Cornell, Lok directs the Rural-Urban Building Innovation Lab. Working with non-standardized material, her research and teaching explore the intersection of technology, novel material methods, and urbanization to develop hybridized design and construction processes.

Sasa Zivkovic is a co-founder at HANNAH, an experimental design practice that utilizes innovative forms of construction to create architectural projects across scales. Zivkovic is also Assistant Professor at Cornell University AAP, where he directs the Robotic Construction Laboratory (RCL), a research group that develops novel robotic construction technology based on sustainable material systems.

Oyster Tiling

Augmented Ecological Topology

Alex Schofield, Megan Considine, Dr. Steven Rumrill

This project supports ecological infrastructure that pro-
motes health and biodiversity, while also contributing value
to human environments. This work ultimately furthers the
concept of not building solely for the binary of human or
natural environment, but in symbiosis of both.

Oyster Tiling is a study that blends material science,
computational design, advanced fabrication, and restoration ecology to produce a tiling substrate for recruitment
of functionally extinct Olympia oysters at Coquille Point on
Yaquina Bay in Oregon. We utilized a design methodology
with careful consideration of the intervention at the scale
of Micro (material), Meso (topological), and Macro (tiling/
deployment).

MICRO - MATERIAL COMPOSITION

Conventional interventions of various intertidal structures
often include use of concrete for its abundance and low
cost. However, concrete often contains basic pH that is
too high for organic life and is carbon intensive in creation.
For this reason, we investigated:

PRODUCTION NOTES

Architect:	Alex Schofield
Restoration Manager:	Megan Considine
Principal Investigator:	Dr. Steven Rumrill
Client:	Oregon Department of Wild Life, on behalf of the oysters
Status:	Ongoing
Site Area:	270,000 sq. ft.
Location:	Coquille Point, Yaquina Bay, Oregon
Date:	2016

1   *Oyster Tiling* made primarily of Calcium Carbonate, laid out to reveal
    patterns and variations of augmented ecological typology for the settlement
    and growth of Olympia oysters

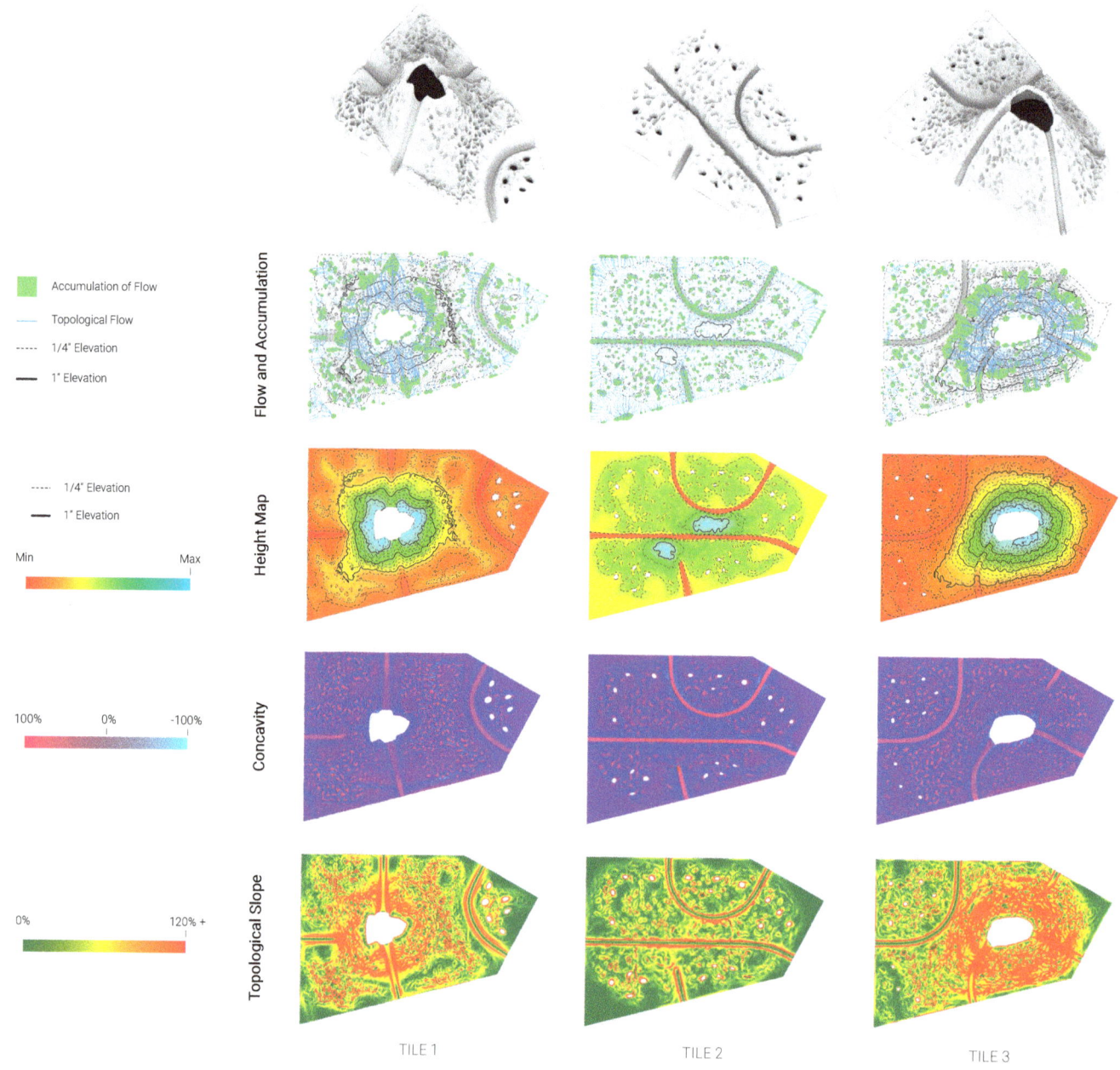

Flow and Accumulation

- Accumulation of Flow
- Topological Flow
- 1/4" Elevation
- 1" Elevation

Height Map

- 1/4" Elevation
- 1" Elevation

Min — Max

Concavity

100% — 0% — -100%

Topological Slope

0% — 120% +

TILE 1   TILE 2   TILE 3

2   Diagrams of topological analysis to predict performance through optimization of form and surface

- Calcium Carbonate, the chemical composition of oyster shells, is the natural settlement substrate of oysters in the wild.
- Magnesium Oxide, as Magnesium Phosphate (MPC), was chosen for its fast hardening, near-neutral pH, and high bending/compressive strength (Yuan et al. 2021).
- Zinc Oxide, as Zinc Phosphate, was similarly chosen for its biocompatibility (commonly used in dental fillings) (Wagh 2016).

MICRO - MATERIAL TESTING

These materials were cast into 4" x 2" x 0.5" tiles in order to test suitability for oyster larvae recruitment. In the wild, Olympia oyster larvae are released into the water column, remain planktonic for several days, and settle onto a substrate in which they become 'oyster spat.' The tiles were placed in a tank within a controlled hatchery setting, introduced with planktonic oyster larvae, and then removed to count the number of oyster spat that settled. A higher spat count was observed in calcium carbonate tiles compared

Oyster Tiling Schofield, Considine, Rumrill

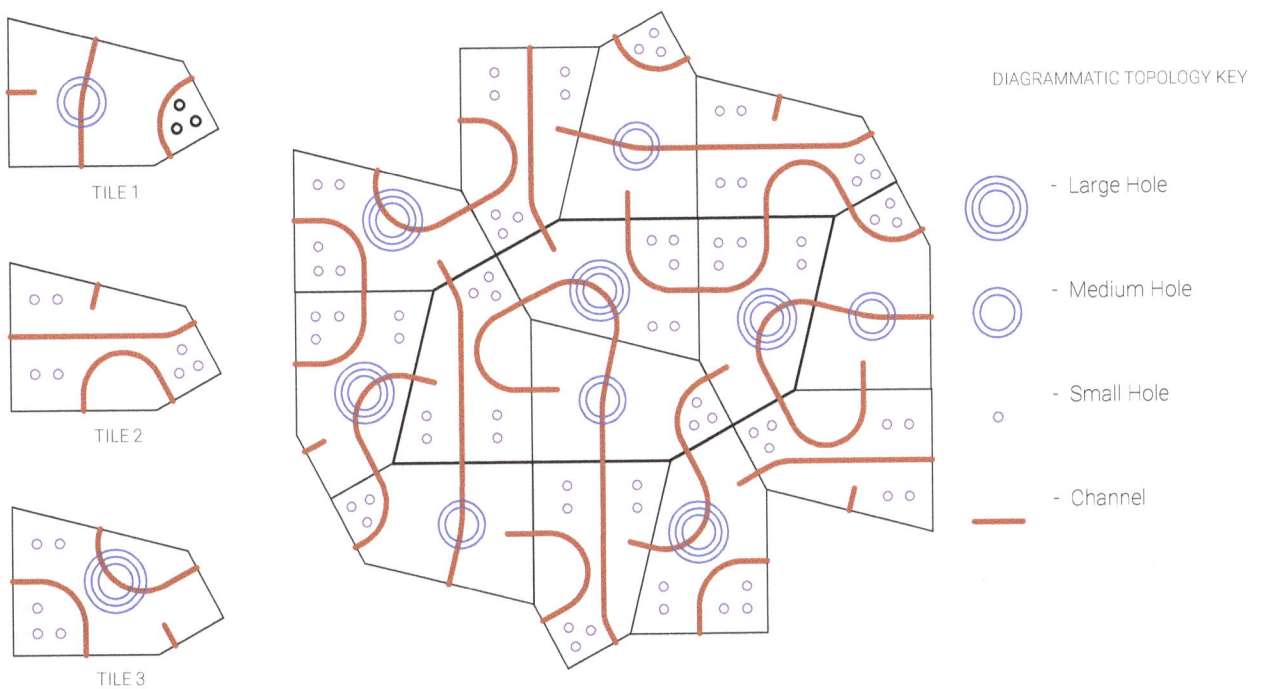

TILE 1

TILE 2

TILE 3

- Large Hole

- Medium Hole

- Small Hole

- Channel

3 Diagrammatic schematics used as visual language of topological elements for ease in communication with marine biologists and team

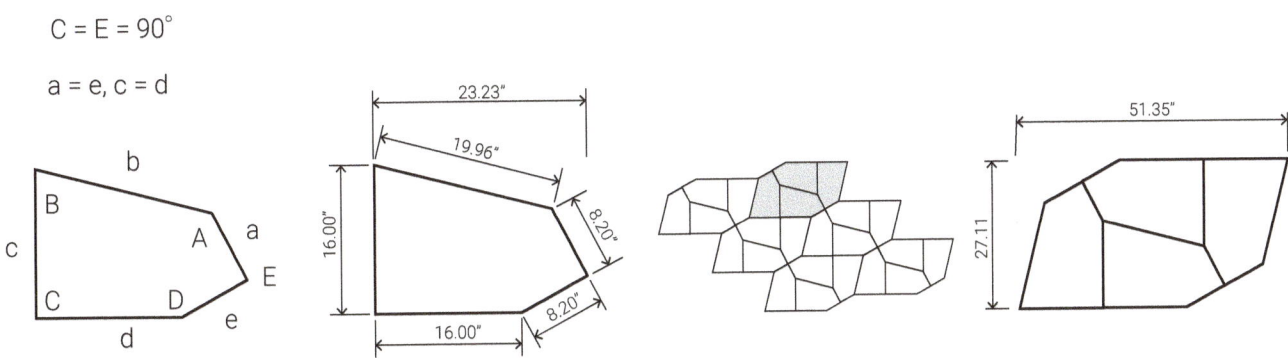

$C = E = 90°$

$a = e, c = d$

b

B

A

a

c

C

E

D

d

e

23.23"

19.96"

16.00"

8.20"

16.00"

8.20"

51.35"

27.11

4 Isohedral tiling "Type 4" (Reinhardt 1918) proof, selected to limit the amount of unique tiles by avoiding symmetry

to all other material compositions confirming that a material similar to their natural settlement substrate would be ideal.

MESO - TOPOLOGY

Spatial observations led us to believe that optimization of topology to mimic similar natural spatial phenomena would be ideal. Identification of three primary topological types assists oysters in growth and maturation; rugosity for increase of surface area as opportunity for settlement, holes for refuge and access through the tiles, and channels for accumulation and water flow with daily tidal changes.

Each typological element was parametrically limited in size so they would be smaller than the oyster's primary predators (e.g., invasive European green crab) yet big enough for the oyster's growth.

MACRO - TILING

Application of various topological elements, in support of Olympia oyster recruitment and growth, required creative consideration of tiling methodologies for deployment in tidal landscape. Isohedral tiling was chosen for a variety of pentagonal units of which "Type 4" (Reinhardt 1918)

5 Arrangement of tiles at site location 1 of 3 selected locations for continued long term observation and study

6 Coquille Point site location at low tide, showing site rugosity and intertidal conditions. The site consists of century-old infill from rocks used as ballast for ships from unknown origins; however, composition of local rock to the area is primarily basalt from historic lava flows.

was selected as it does not require mirroring, reducing the number of unique tiles. The topological features were then applied, similar to Japanese Karakusa tiling methodologies, in a way that no matter how the tile is rotated it aligns and continues the pattern. Three unique tiles were computationally generated and designed, leveraging this tiling methodology, so once deployed there would appear enough variation in a multitude of combinations as seen in the chaotic assembly of the natural world.

DEPLOYMENT

The tiles were rapidly fabricated on site using a 1/2 inch flexible mold liner and plaster mother mold, vastly reducing material usage in mold making. Oyster tiles were of manageable size and weight such that an individual could easily handle a single tile. Due to the funding cycle, tiles were fabricated after the Olympia oyster spawning season, thus only twelve initial tiles were deployed to observe settlement of non-targeted species. All thirty tiles were deployed in July 2022 during the spawning season and will be monitored to observe establishment and growth of Olympia oysters as they settle onto new, augmented, ecological infrastructure.

ACKNOWLEDGMENTS

We would like to express our sincere appreciation to Oregon Oyster Farm for providing Olympia oyster adult broodstock and Whiskey Creek Hatchery (WCH) for settling the Olympia oyster cultch. We would also like to thank WCH for allowing us to conduct the tile experiment in their settlement tanks. This work was supported by the Pacific Marine and Estuarine Fish Habitat Partnership [Grant No: F21AP00858-00 / PMEP: Enhancement of Olympia Oysters to Provide Heterogeneous Habitat for Fish and Invertebrates].

REFERENCES

Hernández, A.B., R.D. Brumbaugh, P. Frederick, et al. 2018. "Restoring the eastern oyster: how much progress has been made in 53 years?" *Frontiers in Ecology and the Environment* 16 (8): 463–471.

Reinhardt, Karl. 1918. *Über die Zerlegung der Ebene in Polygone.* Borna-Leipzig: Noske.

Theuerkauf, Seth J., Russell P. Burke, and Romuald N. Lipcius. 2015. "Settlement, Growth, and Survival of Eastern Oysters on Alternative Reef Substrates" *Journal of Shellfish Research* 34(2): 241–250.

7 "Oyster View" of deployed tiles up close; tiles are placed on top of site condition, often rugose and uneven, where tidal waters may flow over and around

Wagh, Arun S. 2016. "Zinc Phosphate Ceramics." Chap. 10 in *Chemically Bonded Phosphate Ceramics*. 2nd ed. Cambridge, MA: Elsevier. 133–139.

Yuan, Qiang, Zanqun Liu, Keren Zheng, and Cong Ma. 2021. "Inorganic Cementing Materials." Chap. 2 in *Civil Engineering Materials*. Cambridge, MA: Elsevier. 17-57.

IMAGE CREDITS

All images and drawings by the authors.

Alex Schofield is Adjunct Professor of Architecture at California College of the Arts, San Francisco, and collaborator of the Architectural Ecologies Lab. He directs Objects and Ideograms, a design workshop based in Oakland, California. His award-winning work has been exhibited internationally and focuses on material innovations in fabrication, computation, and ecology. Alex is a graduate of University of California, Berkeley, College of Environmental Design.

Megan Considine is a Restorative Aquaculture Fellow for The Nature Conservancy and Oregon Sea Grant and a Project Facilitator for Mujeres de Islas. She is interested in community based regenerative aquaculture for its potential to provide food sovereignty, while maintaining and in some cases even improving the environment through key ecosystem services. Megan also has experience in native oyster restoration work on the East and West Coasts of the United States. At the time of this project Megan was completing her master's degree in marine resource management at Oregon State University.

Dr. Steven Rumrill is the Shellfish Program Leader for the Oregon Department of Fish and Wildlife. His primary research interests lie in the interdisciplinary links between invertebrate reproductive biology, life-history ecology, and ecosystem functions in the marine and estuarine environments. Dr. Rumrill has worked with a variety of organisms in several marine ecosystems across the United States, including native Olympia oyster restoration work within Oregon's semi-enclosed estuaries.

Tortuca

An Ultra-Thin Funicular Hollow Glass Bridge

Yao Lu, Alireza Seyedahmadian, Philipp Amir Chhadeh, Matthew Cregan, Mohammad Bolhassani, Jens Schneider, Joseph Robert Yost, Gareth Brennan, Masoud Akbarzadeh

Designed with Polyhedral Graphic Statics (PGS), a geometry-based structural form-finding method, Tortuca presents an efficient and innovative structural system constructed by the dry assembly of thirteen hollow glass units (HGU). It also proposes a new language for glass that is carefully treated, structurally informed, fabrication-aware, and environmentally responsible. Each HGU of Tortuca is made of 1 cm (3/8 inch) glass deck plates and 2 cm (0.7 inch) acrylic side plates precisely cut with 5-axis abrasive waterjet cutting and CNC milling to match the structural geometry. The structure spans 3.2 m (10.5 ft) with a mass of only 250 kg (550 lbs), where the float glass is the primary load-bearing material. Thanks to the efficiency and light weight of the construction system, a single person can assemble and disassemble the structure without needing a crane or additional labor. Moreover, this research explores the potential of using an extremely delicate material such as float glass for the primary structural system to encourage minimizing the material and energy demands in buildings and infrastructural projects. Additionally, it shows how utilizing the material in its purest format could simplify the recycling process after the life cycle of the structure has ended. Also, this research project is achieved by collaboration across different institutions, from design to engineering, from theoretical to practical, and from academia to industry. We appreciate the value of breaking disciplinary boundaries and joining forces from multiple fields.

STRUCTURAL FORM-FINDING

The generation of the base geometry (Figure 3) is achieved through PolyFrame (Nejur and Akbarzadeh 2021), a Rhino plug-in that implements PGS. The generated form diagram with thirteen polyhedral cells is then optimized regarding global dimensions, individual edge lengths, and face angles, to better satisfy the fabrication constraints and increase the ease of the subsequent fabrication process. As a result, the bridge dimension is set to 3.2 m(L) × 1.3 m(W) ×0.55 m(H), the thicknesses of the polyhedral cells are set to around 100 mm, and each cell is constrained to a size that one person can handle during the construction process.

MATERIALIZATION

After obtaining the base geometry of the bridge, each polyhedral cell is materialized as one HGU. The HGU details and

PRODUCTION NOTES

Architect:	Yao Lu, Masoud Akbarzadeh
Status:	Completed
Site Area:	50 sq. ft.
Date:	2022

1   Side view of the assembled hollow glass bridge structure

2 The assembled hollow glass bridge structure

the steel abutments that hold all HGUs are devised based on the material selection and fabrication constraints. Each of the two top and bottom faces of every polyhedral cell is materialized as a glass deck plate using 9.5 mm annealed glass. For the smaller side faces, they are materialized as either 9.5 mm thick glass side plates or 21 mm thick acrylic side plates, depending on whether they need to accommodate the connection mechanisms with the neighbor HGUs (Figure 4). Improved from a precedent (Lu et al. 2021), the connection mechanism between neighboring HGUs contains two pocket channels on the pair of facing side plates and a locking strip that has a butterfly shape section profile. Between the neighboring HGU deck plates and between the HGUs and steel abutments, Surlyn sheets cut with a 3-axis CNC router are placed and used as the interface material preventing direct glass-to-glass, and glass-to-steel contact.

NUMERICAL SIMULATION

Before HGU fabrication and bridge assembly, a static numerical finite element analysis is conducted using ANSYS, which follows the previously established analysis sequence as explained by Yost et al. (2022). The result shows that the bridge successfully sustains its self-weight.

FABRICATION AND ASSEMBLY

The glass parts are cut using 5-axis abrasive waterjet cutter (Figure 5), and the acrylic parts are milled using 5-axis CNC router (Figure 6). The assembly process consists first of the construction of the individual HGUs (Figure 7) followed by the assembly of the entire bridge (Figure 9). The heaviest HGU of the bridge weighs about 23.3 kg, meaning that the assembly process can be handled by one person without any heavy construction machinery. Moreover, most material can be easily dismantled and recycled at the end of its life cycle as a consequence of the dry assembly process.

ACKNOWLEDGMENTS

This project was supported by University of Pennsylvania Research Foundation Grant (URF), National Science Foundation CAREER AWARD (NSF CAREER-1944691-CMMI), and the National Science Foundation Future Eco Manufacturing Research Grant (NSF, FMRG-CMMI 2037097) to Dr. Masoud Akbarzadeh. It was also supported by Villanova University Summer Grant Program (USG) to Dr. Joseph Yost. The multi-axis milling, metalwork, and other facilities were supported by Eventscape NY.

REFERENCES

Lu, Y., A. Seyedahmadian, P.A. Chhadeh, M. Cregan, M. Bolhassani, J. Schneider, J.R. Yost, G. Brennan, and M. Akbarzadeh. 2022. "Funicular

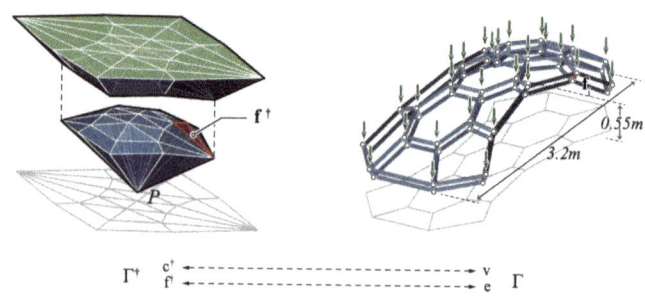

3 Form-finding using Polyhedral Graphic Statics (PGS)

Tortuca Lu, Seyedahmadian, Chhadeh, Cregan, Bolhassani, Schneider, Yost, Brennan, Akbarzadeh

4 Close-up of the butterfly connection mechanism

5 5-axis abrasive waterjet cutting for glass fabrication

6 5-axis CNC milling for acrylic fabrication

Hybrids & Haecceities 169

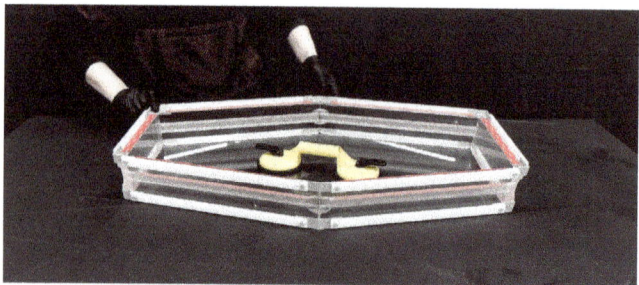

7   Construction of a typical hollow glass unit

8   The locking strips can be easily inserted and removed during assembly

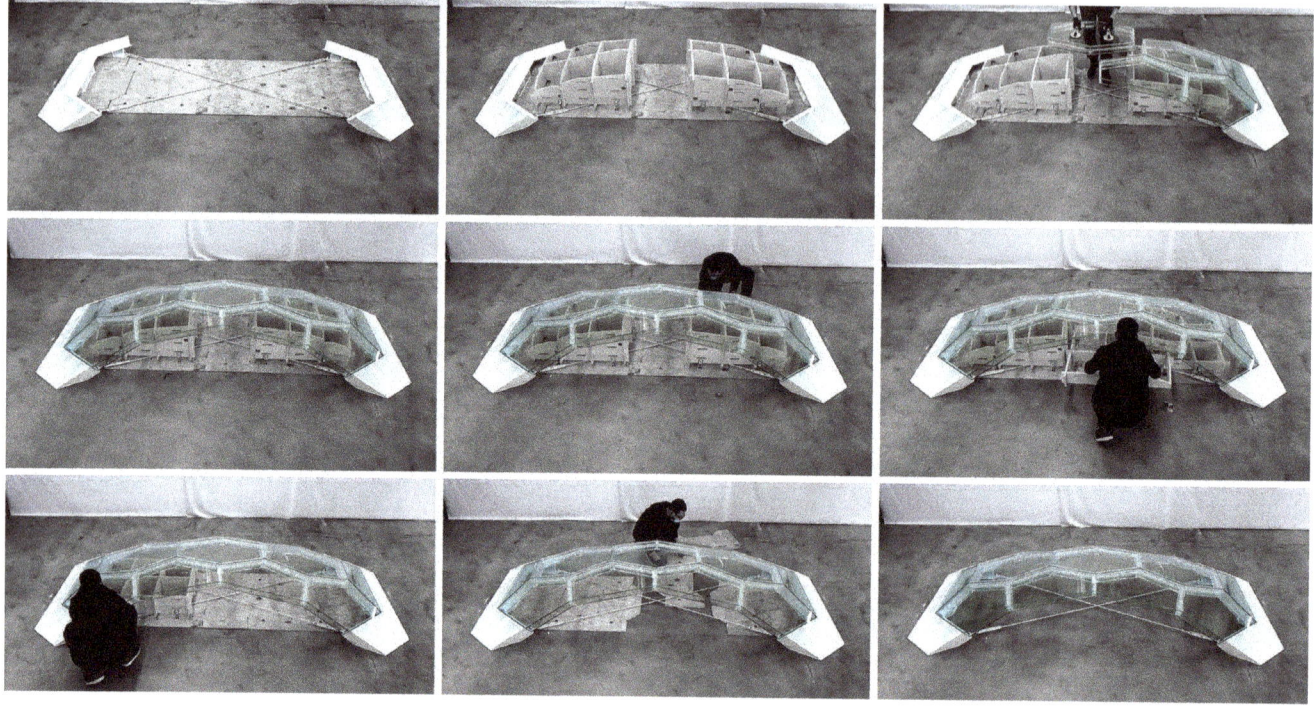

9   The modular assembly process of the structure can be handled by one person thanks to the lightweight of the hollow glass units

Glass Bridge Prototype: Design Optimization, Fabrication, and Assembly Challenges." *Glass Structures & Engineering* 7 (2): 319–330.

Lu, Y., M. Cregan, P.A. Chhadeh, A. Seyedahmadian, M. Bolhassani, J. Schneider, J. Yost, and M. Akbarzadeh. 2021. "All Glass, Compression-Dominant Polyhedral Bridge Prototype: Form-Finding and Fabrication." In *Inspiring the Next Generation: Proceedings of the 7th International Conference on Spatial Structures and the Annual Symposium of the IASS.* Surrey, UK. 326–36.

Nejur, A., and M. Akbarzadeh. 2021. "PolyFrame, Efficient Computation for 3D Graphic Statics." *Computer-Aided Design* 134 (May): 103003.

Yost, J.R., M. Bolhassani, P.A. Chhadeh, L. Ryan, J. Schneider, and M. Akbarzadeh. 2022. "Mechanical Performance of Polyhedral Hollow Glass Units under Compression." *Engineering Structures* 254 (March): 113730.

IMAGE CREDITS

All other drawings and images by the authors.

Yao Lu is a PhD student of Architecture at the Polyhedral Structures Laboratory (PSL), University of Pennsylvania. He is a design researcher with interests in generative design, computational techniques, and digital fabrication. He holds a Master of Science in Matter Design Computation degree from Cornell University, and a Master's in Architecture and a Bachelor's in Engineering from Tongji University.

Alireza Seyedahmadian is a Senior Design Engineer at Eventscape, where he leads the Advanced Manufacturing Department in their Long Island City facility. Ali is an artist, designer, and maker with a background in Architecture and Digital + Robotic Fabrication. He holds a Master's in Architecture from the University of Michigan and has worked primarily in leading custom fabrication companies for the world of art and architecture.

Philipp Amir Chhadeh is a PhD student at the Institute of Structural Mechanics and Design, Technische Universität Darmstadt. He also works as a Research Assistant at Glass Competence Center and Generative Design Lab.

Tortuca Lu, Seyedahmadian, Chhadeh, Cregan, Bolhassani, Schneider, Yost, Brennan, Akbarzadeh

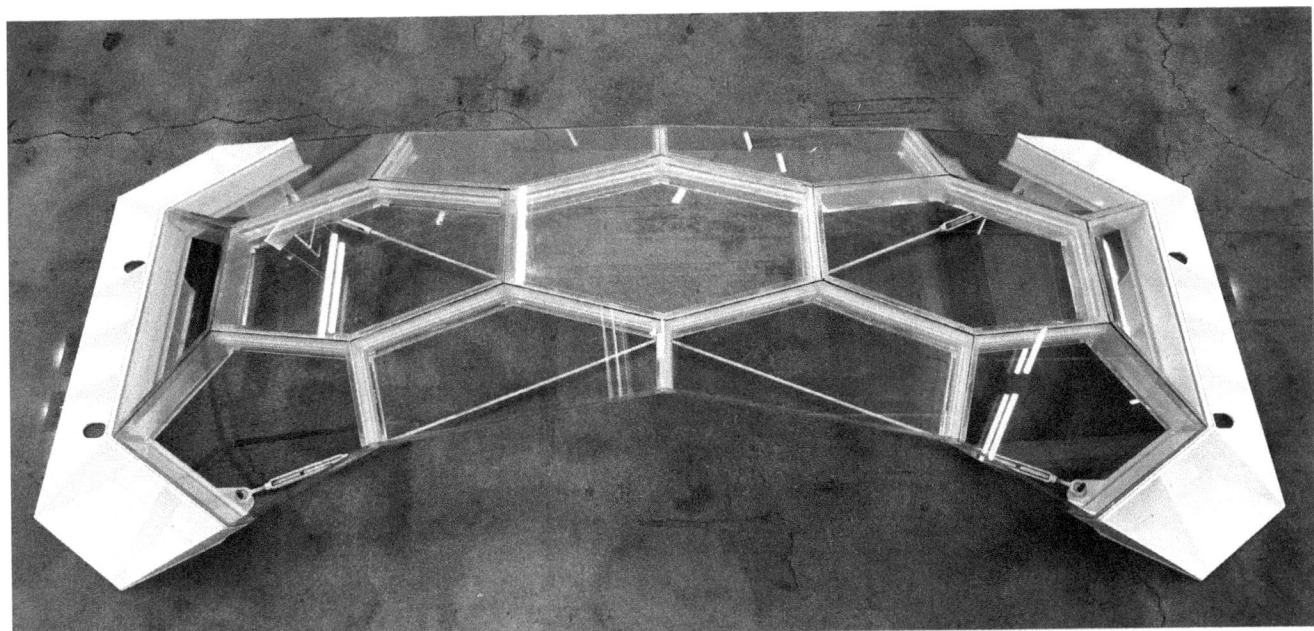

10 Aerial view showing both the compression and tension members

11 View looking up at the steel abutment

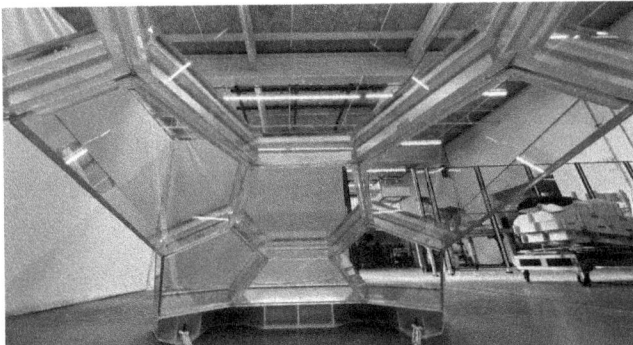

12 View from under the glass bridge structure

Matthew Cregan received a Master of Science in Civil Engineering with a structural concentration from Villanova University in 2022. He is currently employed as a bridge engineer with Hardetsy & Hanover in New York City.

Mohammad Bolhassani is Assistant Professor at the Bernard & Anne Spitzer School of Architecture, the City College of New York. His experience covers computational and experimental analysis, design, retrofit, and rehabilitation of structures.

Jens Schneider is Professor at the Institute of Structural Mechanics and Design, Technische Universität Darmstadt. His research interests include analysis of complex structural systems and innovative materials in construction engineering.

Joseph Robert Yost is Professor in the Department of Civil and Environmental Engineering at Villanova University. His research interests include investigation of novel materials and innovative structural systems for application in civil infrastructure.

Gareth Brennan is the President and Founder of Eventscape, an elite art and architectural fabrication company. Founded in 1993, Eventscape is an internationally recognized, award-winning company with 150,000 sq. ft. of state-of-the-art manufacturing facilities in Toronto and a 20,000-sq. ft. state-of-the-art studio in Long Island City, supported by a talented team of two hundred architects, designers, engineers, and craftsmen. The company's projects have won over 150 major design industry awards to date. In 2009, Gareth was named Canadian CEO of the Year for best design strategy at the annual Design Exchange Awards.

Masoud Akbarzadeh is Assistant Professor of Architecture in Structures and Advanced Technologies and the Director of the Polyhedral Structures Laboratory (PSL) at University of Pennsylvania. His main research topic is Three-Dimensional Graphic Statics, a novel geometric method of structural design in three dimensions. In 2020, he received National Science Foundation CAREER Award to extend the methods of 3D/Polyhedral Graphic Statics for Education, Design, and Optimization of High-Performance Structures.

Robotically Bent Spatial Metal Knots

Re-Interpreting Spatial Knotting Through Robotic Tube Bending

Lee-Su Huang, Gregory Thomas Spaw

The studies presented in this project employ workflows that enable multi-planar robotic bending of metal tubes with high accuracy and repeatability using a 6-axis industrial robot with a custom end-effector and external 2-axis positioner (Huang, Spaw, and Kalo 2022). This enables quick fabrication and assembly of complex spatial tubular configurations without the need for jigs or falsework support. The method makes possible the accurate fabrication of complex, closed polyline curve loop configurations that are topologically similar to knots in mathematical knot theory (Sosinskiĭ and Bronislavovich 1997).

As a test case to verify and demonstrate the accuracy and tolerances of the workflow, the project draws inspiration from cultural crafts and spatial knotting traditions, reinterpreting them as tubular constructs. While knots are commonly understood as dense and compressed, decorative Chinese knots are planar, often constructed out of a single continuous string, and exhibit space in between, with the knot body made of two layers of cord sandwiching an empty space (Chen 2007). Therefore, this project takes on the notion of weaving or knotting tubes as a continuous material trajectory (as opposed to disjointed), and eschews welding in favor of mechanical fasteners that can be assembled or disassembled with ease. The resultant geometries create inherent opportunities for joints, self-support, and crossings that reinforce structural behavior similar to reciprocal structures while serving as a more rigid synthetic counterpart to the softer organic nature of the knots.

PRODUCTION NOTES

Hardware:	Kuka KR 125/3
	DKP-400 Positioner
	Custom gripper
Software:	Rhino+Grasshopper
	Lunchbox for GH
	KukaIPRC
Pipe Dia:	16/14 mm OD/ID
Location:	AUS CAAD Labs
Date:	2022

1   Physical studies of spatial knotting fabricated through robotic tube bending

Hybrids & Haecceities 173

2 Large-scale installation showing bundling detail

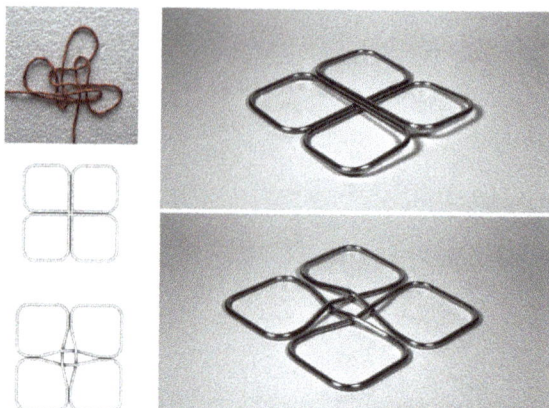

3 A physical clover leaf knot reference and two minimal depth studies with corresponding top views

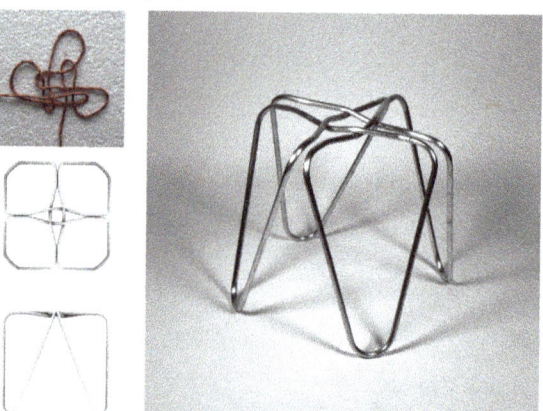

4 The physical clover leaf knot reference and an expanded depth study with corresponding top and elevational views

PROCESS

As an initial step, abstracted polyline traces of the original knot diagrams are drawn in 2D, inserting control points at intersections that correlate to locations where the polyline needs to deflect up or down to avoid self-intersecting. A series of geometric studies develop from these initial polylines, exploring different configurations of the knots in three dimensions, maintaining continuity and avoiding self-intersections. Five initial knot diagrams were chosen for reinterpretation, and each was developed into a series of digitally modeled knot variations, experimenting with spacing, crossings, and viable geometries to understand the rulesets and constraints in play for each knot (Figures 3-8). This process is conceptually similar to how mathematicians studying knot theory use lattice topology models to represent possible relationships within the same knot (Hayes 1997). The design workflow of these tubular reinterpretations of knots requires careful and deliberate manipulation of closed polyline loops. The segmentation strategy for the loop is instrumental to producing geometry that is possible to fabricate, avoiding self-intersection or collision constraints with the physical workspace.

RESULTS

The geometric information is parsed and sequenced into Kuka KRL code from Kuka|PRC for the robot movement simulation, collision detection, as well as external axis instructions for the turntable and gripper to actuate in choreography with the robot movement. In total, seven physical prototypes were chosen to be fabricated from the digital tests, dimensionally ranging from 50-90 cm wide and up to 80 cm tall (Figure 1). The flatter 2.5D interpretations were more difficult to execute due to the very minimal bend angles and limited linear distance for the tubes to clear their intersection. A primary limitation is the 40 mm minimum distance between adjacent bends, due to the spacing of the bending pin, which forces more separation in order to navigate crossings with enough clearance. The largest experiment is a 2.3-meter tall construct consisting of two

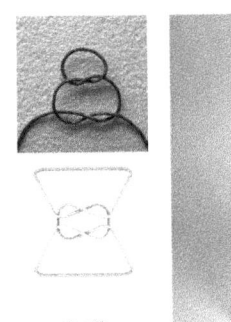
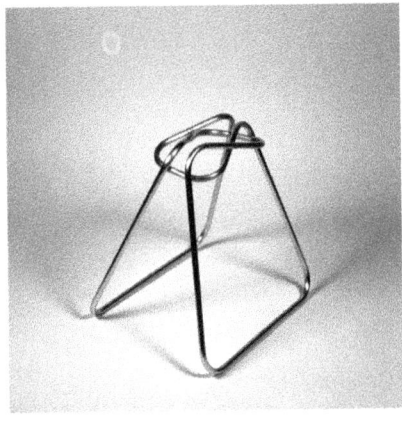

5   A physical flat knot reference and an expanded depth study with corresponding top and elevational views

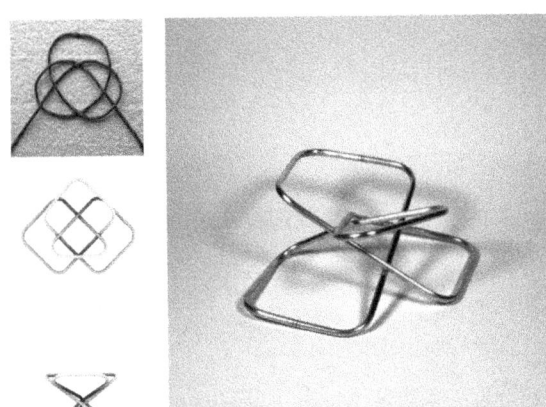

7   A constellation knot reference and an expanded depth study with corresponding top and elevational views

6   A physical double coin knot reference and an expanded depth study with corresponding top and elevational views

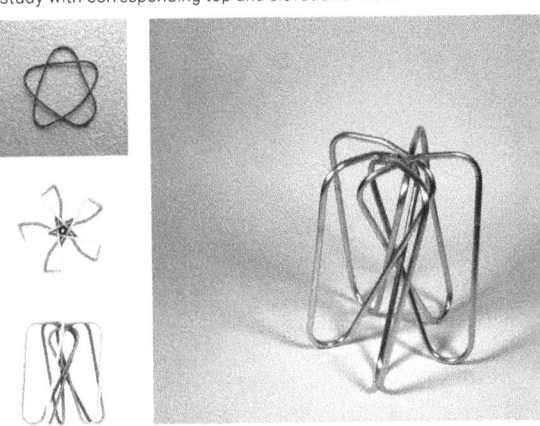

8   A 51 knot reference and an expanded depth study with diagram and corresponding top and elevational views

interwoven color-coded continuous loops that are folded upon themselves in pentagonal rotational symmetry, with each side consisting of seven unique segments, for a total of 35 tube segments that have slight geometric variations in the woven center bundle to avoid self-intersection (Figures 2, 11).

FUTURE WORK

While the project focuses on self-contained loops to stay within the spirit of the knot inspiration, there are clear future implications for employing multiple elements working in bundles or weaves, as demonstrated in the test prototype shown (Figure 9). These elements may cross multiple structural and volumetric cells, employing a variety of joint conditions (crossing, end-to-end, parallel) and tectonic expressions. The development of joints or fittings that allow for connections to planar elements is also underway (Figure 10). The geometrical constraints of the workflow, including bend radius, minimum distance between bends, and self-intersection or collision constraints with the

physical workspace and equipment should be codified into morphological rulesets that can act as design guidelines. As part of a series of studies intended to gradually scale up in size and complexity, the project reveals possibilities for bundles, knots, and interweaving that represent a paradigm shift in how tubular structures might be designed and fabricated in the future, uncovering a novel material expression that is simultaneously malleable and rigid.

ACKNOWLEDGMENTS

This project was made possible by an American University of Sharjah Design Build Initiative Skillset Development Grant and a Faculty Research Grant.

The facilities and personnel at the American University of Sharjah CAAD Fabrication Labs were invaluable to the development process of the project, in particular the assistance of Director of CAAD Labs and Associate Professor Ammar Kalo.

REFERENCES

Braumann, Johannes. *Kukalprc*. V.2020-01-24. Association for Robots in Architecture. PC. 2020.

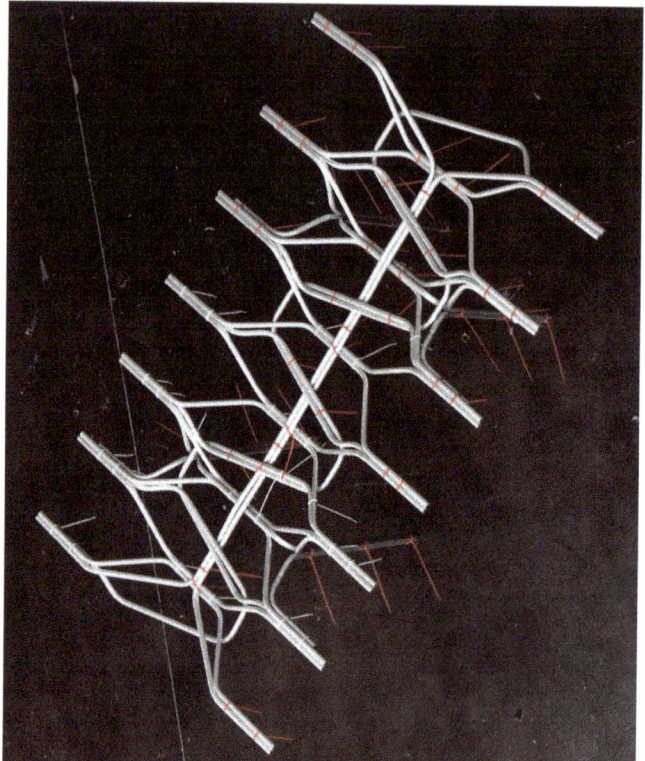

9 Test prototype utilizing bundling, weaving, and knotting of multiple tube trajectories

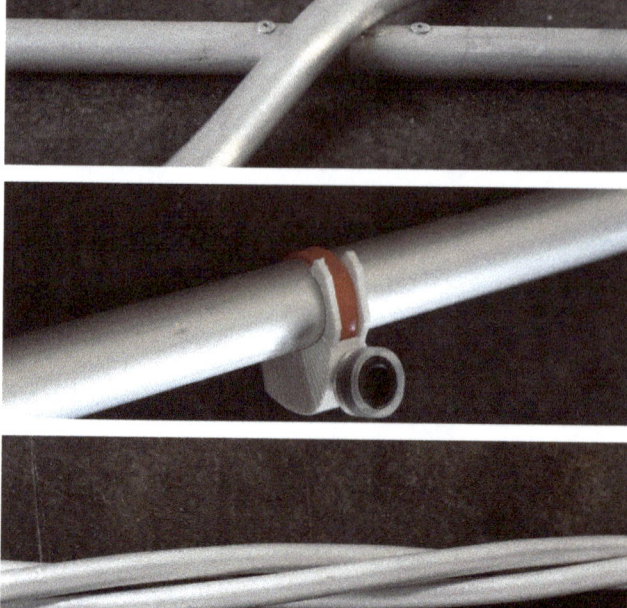

10 Crossing, prosthetic, and twisting local connection studies

Chen, Lydia. 2007. *The Complete Book of Chinese Knotting: A Compendium of Techniques and Variations*. Singapore: Tuttle Publishing.

Hayes, Brian. 1997. "Computing Science: Square Knots." *American Scientist* 85 (6): 506–10. http://www.jstor.org/stable/27856883.

Huang, Lee-Su, Gregory Spaw, and Ammar Kalo, "Multiplanar Robotic Tube Bending," *International Conference for the Association for Computer-Aided Architectural Design Research in Asia CAADRIA 2022: Post-Carbon*. Sydney, Australia, April 9-15, 2022.

McNeel. *Rhinoceros+Grasshopper*. V.7.7.21160.5001. Robert McNeel & Associates. PC. 2020.

Miller, Nathan. *Lunchbox for Grasshopper*. V.2020.11.2.0. Proving Ground LLC. PC. 2020.

Sosinskiĭ, A.B., V.V. Prasolov. 1997. *Knots, Links, Braids and 3-manifolds: An Introduction to the New Invariants in Low-dimensional Topology*. Providence, RI: American Mathematical Society.

IMAGE CREDITS

All drawings and images by the authors.

Lee-Su Huang received his Bachelor of Architecture from Feng-Chia University in Taiwan and his Master in Architecture degree from Harvard University's Graduate School of Design. He has practiced in Taiwan with various firms and in the United States with Preston Scott Cohen Inc. in Cambridge, and with LASSA Architects in Seoul, Korea. As co-founder and principal of SHO, his research and practice centers on digital design and fabrication methodology, parametric design optimization strategies, as well as kinetic/interactive architectural prototypes. Lee-Su is currently Instructional Assistant Professor at the University of Florida's School of Architecture, teaching graduate and undergraduate design studios as well as foundation and advanced digital media and parametric modeling courses.

Gregory Thomas Spaw is Associate Professor at the American University of Sharjah in the United Arab Emirates. He has previously held the Ann Kalla Assistant Professorship at Carnegie Mellon University and taught undergraduate and graduate studios, seminars and electives at the University of Tennessee. Concurrent with his academic engagement, Spaw is a principal of SHO, a design collaborative that straddles the territories of teaching, research and practice. His previous professional experience includes work with the award-winning offices of Bohlin Cywinski Jackson, Preston Scott Cohen Inc., and Asymptote. His scholarly pursuits incorporate digital visualization, harnessing parametric workflows, intelligent material fabrication, and responsive environment design.

11 Installation utilizing pentagonal rotational symmetry and similar principles as the metal knots, consisting of two interwoven, continuous curves designed to interlock and intertwine to offer opportunities for joinery and structural support

Long Range

Shaping Glass for Acoustic and Optic Performance

Catie Newell, Zackery Belanger, Wes McGee

Long Range is a surface shaped to expose the intrinsic acoustic properties of glass, exhibited as gradients of acoustic behavior. Moving from flat panels at one end to deeply slumped and perforated components at the other, the glass reveals its acoustic properties, ranging from reflection, diffusion, absorption, and transmission. The intention is to merge optical and acoustical performance intrinsically into a surface, and offer an alternative to acoustic treatment by calibrating material geometry and sound.

Each component of *Long Range* was slumped from an acoustically-reflective hexagonal blank. Flat in its initial state, this surface condition is the most ubiquitous state of glass in the built environment. By starting with the same condition, every pane moves through a spectrum of geometric manipulation in the fabrication process, gradually altering how that region

PRODUCTION NOTES

Principal Investigators:	Catie Newell, Zackery Belanger, Wes McGee
Dimensions:	25'-4" L x 6'-6" H
Materials:	Guardian Glass 4mm PrivaGuard®, steel cables, and connectors

1   *Long Range* is a glass surface shaped to exhibit acoustic gradients

Hybrids & Haecceities

2   The components of *Long Range* are designed to be "seen" by sound as a single surface that increases in geometric complexity along its length

of the surface interacts with sound waves. Once slumped, the pieces were paired face-to-face along their edges to make glass bubbles. The two-layer system introduces an additional set of variables, including: air volume; the sidedness or lack of perforations (either both panes perforated, one or the other perforated, or neither pane perforated); concavity, convexity, or flatness; continuity of perforation pattern between neighboring panes; and relationship of surface curvature across panels. The components of *Long Range* are made with Guardian Glass 4 mm PrivaGuard®. This dark green tinted float glass provides a visual register for the shifting geometric details of the surface including alignments, overlaps, and varied surface curvature.

Long Range is the result of multiple phases of research. The first phase focused on material and process, resulting in a fabrication system for the controlled and predictable slumping of glass panels. This phase utilized a variable auxetic cut pattern, which extended the range of possible shapes and introduced openings into the system. The second phase used wave-based computational simulations to study the acoustic behavior of the components in aggregate, up to room-scale. Specifically, the simulations provided a way to assess the diffusive behavior of the slumped forms, which is a historically-neglected and challenging region of acoustic design. The third phase strategized surface dimensions, acoustical behavior extents, and system of assembly. Custom hardware and tension cables were constructed to suspend the glass bubbles from inside of the glass forms. This allowed for glass to be the dominant material for maximum visual transparency and lightness of the system.

The most challenging acoustic regime for glass is absorption, which was achieved using an effect known as Helmholtz resonance. The auxetic patterns open as the glass material

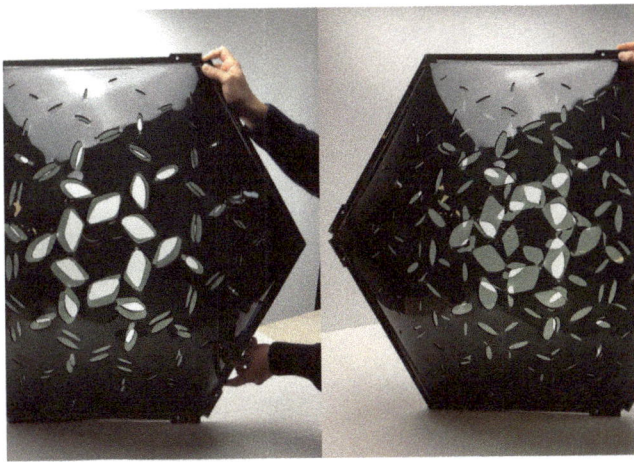

4 Slumped glass pairings yield visual and acoustic complexity

3 Module depths were determined by the wavelengths of the human voice

5 A single flower from *Long Range* undergoing acoustic testing for absorption

unwinds—in a restrained and controlled way due to intentional pattern-fading. This allows the balance of open area and internal air volume to be tuned to absorb particular frequencies. Standardized laboratory testing confirmed the presence of absorption. Three "flower" sections were removed from *Long Range* for these tests, and a fourth benchmark condition of completely flat panels was also included. Tests were conducted at Riverbank Acoustical Laboratories in Geneva, Illinois per ASTM C423 Standard Test Method for Sound Absorption and Sound Absorption Coefficients by the Reverberation Room Method.

All flowers exhibited some measurable absorption, but the more severely-shaped flowers exhibited higher absorption, as hoped and anticipated in the design. The measurable progression of absorption demonstrates the versatility of glass as an acoustic material, without the need for fiberglass. This result also indicates the intrinsic acoustic

potential for glass and acts specifically as evidence against the misconception of glass as a poor acoustic material. *Long Range* shows a continuity between the severity of form, the presence of openings, and acoustic dissipation, indicating that traditional realms of acoustic behavior may be part of a single continuum (Belanger 2021), and that optically-designed acoustic systems are possible.

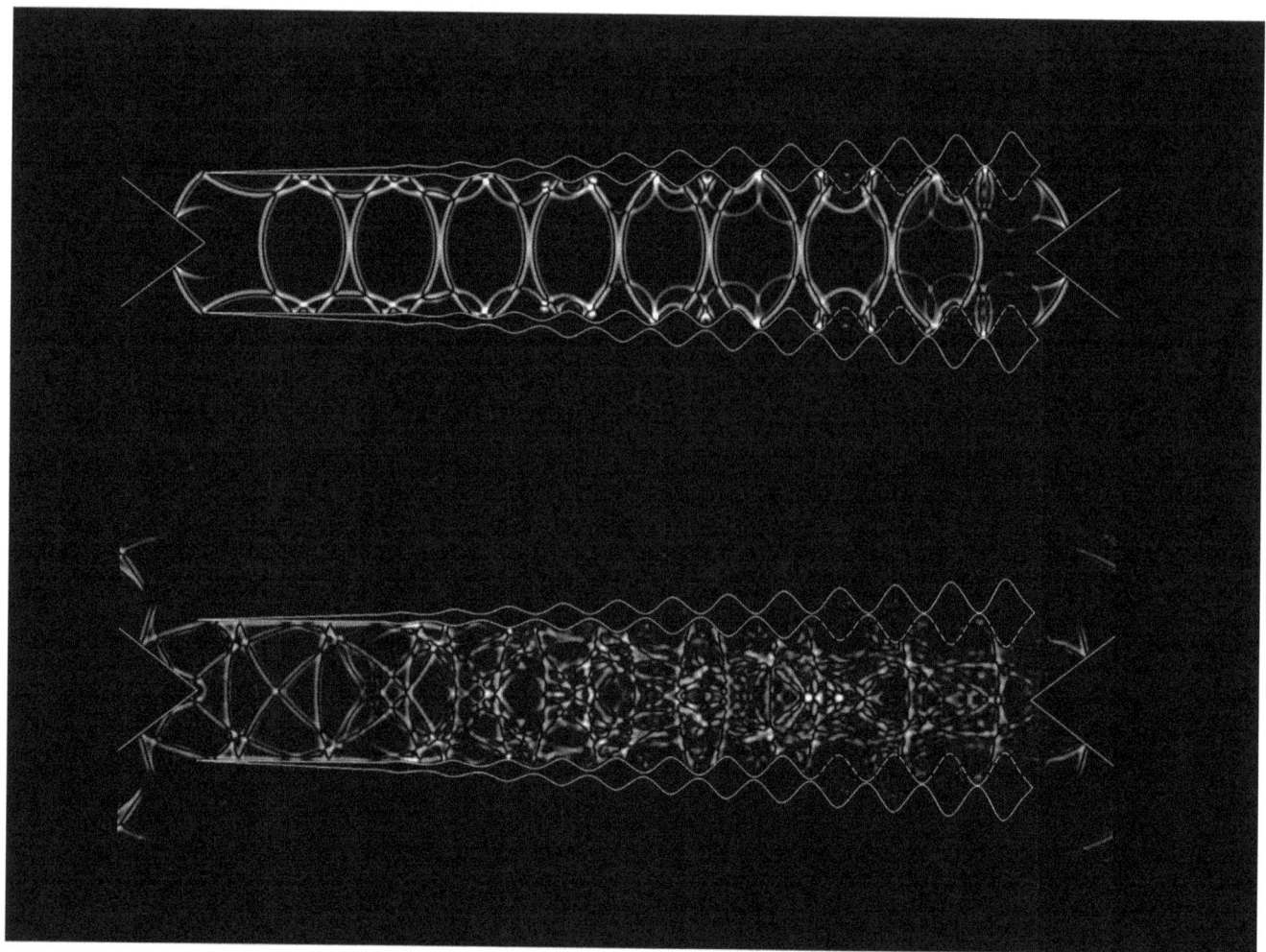

6   Finite difference time domain simulations were conducted to explore surface shaping and the feasibility of an acoustic gradient

ACKNOWLEDGMENTS

This work was funded by Guardian Glass, the University of Michigan Taubman School of Architecture, and Arcgeometer.

Glass was provided by Guardian Glass.

Principal Investigators: Catie Newell, Zackery Belanger, Wes McGee

Fabrication and installation team members include:

Project Leads: Misri Patel, Oliver Popadich

Project Team: Elizabeth Teret, Dan Tish, Maryam Alhajri, Ryan Craney, Amin Aghagholizadeh, Isabelle Leysens, Kelly Gregory

Installation Team: Charlie O'Geen, Mehdi Shirvani, Mackenzie Bruce, Laurin Aman, Jessica Sato.

REFERENCES

Belanger, Zackery. 2021. *Acoustic Ornament*. Detroit: Arcgeometer.

Belanger, Zackery, Wes McGee, and Catie Newell. 2018. "Slumped Glass: Auxetics and Acoustics." In *ACADIA '18: Recalibration, On Imprecision and Infidelity; Proceedings of the 38th Annual Conference for Computer Aided Design in Architecture*. Mexico City, Mexico. 244–249.

Newell, Catie, Zackery Belanger, Wes McGee, and Misri Patel. 2020 "Unknowing." In *ACADIA '20: Distributed Proximites; Proceeding of the 40th Annual Conference of the Association for Computer Aided Design in Architecture*. Online and Globa. 254–257.

7   The hanging structure of *Long Range* consists of tensioned cables that allow the glass surface to dominate the visual and the acoustic

Catie Newell is an Associate Professor of Architecture at the Taubman College of Architecture and Urban Planning at the University of Michigan. and the founding principal of the architecture and research practice Alibi Studio. Newell's work and research captures spaces and material effects, focusing on the development of atmospheres through the exploration of textures, volumes, and the effects of light or lack thereof.

Zackery Belanger is the founder and director of Arcgeometer, a Detroit-based studio that specializes in shaping objects, surfaces, and rooms for acoustic performance. His work pursues the intrinsic acoustic potential of materials and geometry, and aims for the ultimate merging of acoustics into architecture.

Wes McGee is Associate Professor in Architecture and Director of the Fabrication and Robotics Lab at the University of Michigan Taubman College of Architecture and Urban Planning and Co-founder and Partner at Matter Design. McGee's work has been recognized with awards such as the Architectural League Prize for Young Architects & Designers and the ACADIA Award for Innovative Research. His research revolves around the interrogation of design and material production in architecture, with the goal of developing new connections between design, engineering, materials, and manufacturing processes as they relate to the built environment.

Nano.Web.Arch

Nanofibrous Structures Applications in Architecture

Jan Koníček, Miloš Florián, Klára Masnicová, Pavel Pokorný

The *Nano.Web.Arch* project describes new possibilities of using polymers and nanofibrous textiles and their application in architecture. The main ambition of the research was to create an architectural form with the added value of the given materials, such as water retention, sorption, and gradual drying. Such structures or façade elements could contribute to the solution of global problems associated with the decrease of usable and potable water on our planet.

Our inspiration was the construction of webs, the spider colony *Agelena consociata* (Finsterwalderf 2011). Due to its high specific surface, this natural material can retain water from air humidity in the web (Riechert et al. 1986). The project sought similar materials that could replace spider silk and "spider workers" to build the structure (Figure 2).

The research relates to the concept of fog harvesting. Fog harvesting uses textile materials to capture water from atmospheric humidity, and these structures trap moisture on the knitted fabric (Olivier 2014). The advantage of nano-materials compared to knitted fabrics is a higher specific

PRODUCTION NOTES

Architect:	Jan Koníček
Status:	Built
Site Area:	43.06 sq. ft.
Location:	Liberec, Czech Republic
Date:	2021

1 The fundamental idea of the project and the principal idea for new architecture material (Jan Koníček, 2021)

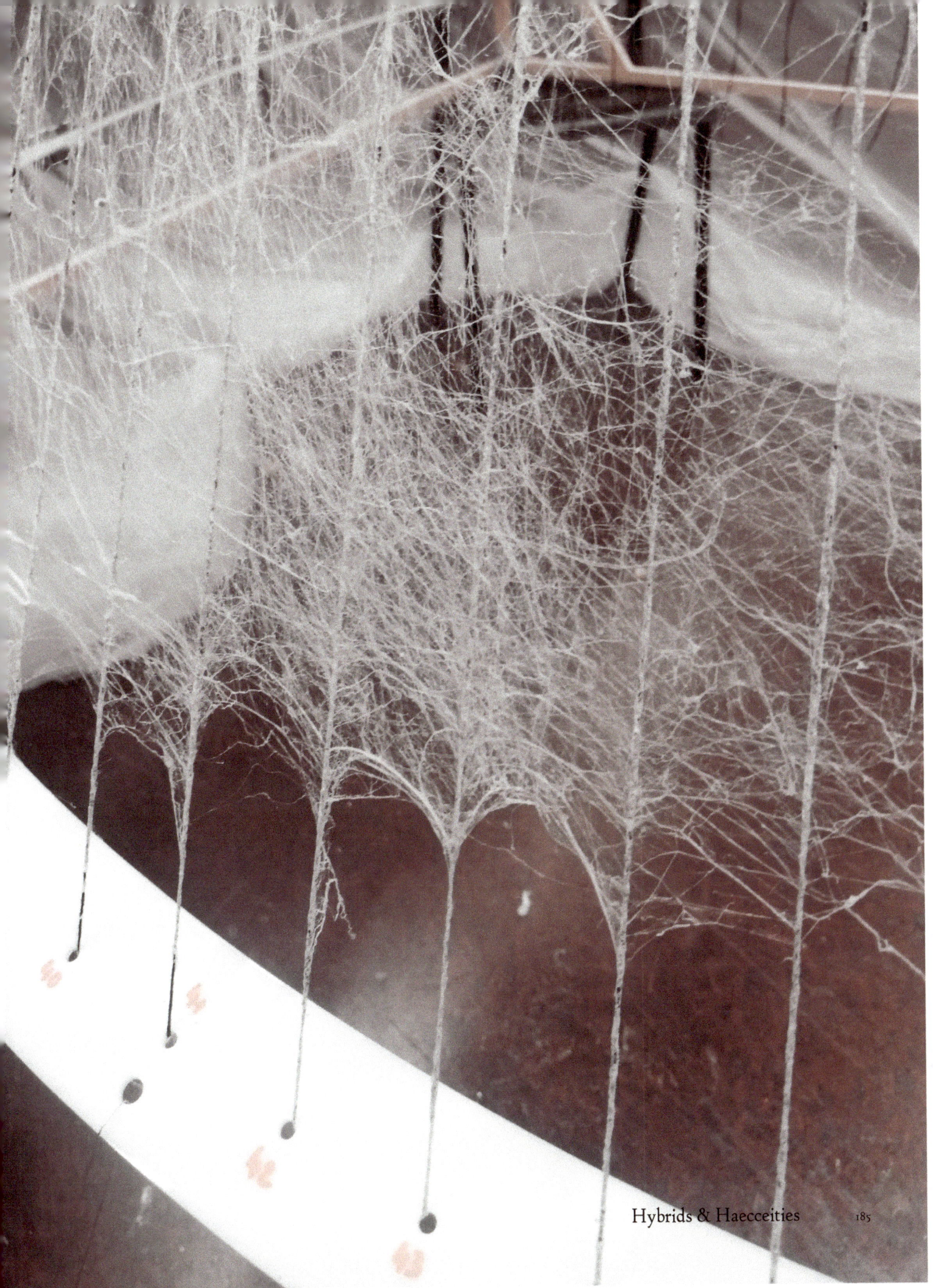

2   Inspiration by spider webs (Anna Machátová, 2021)

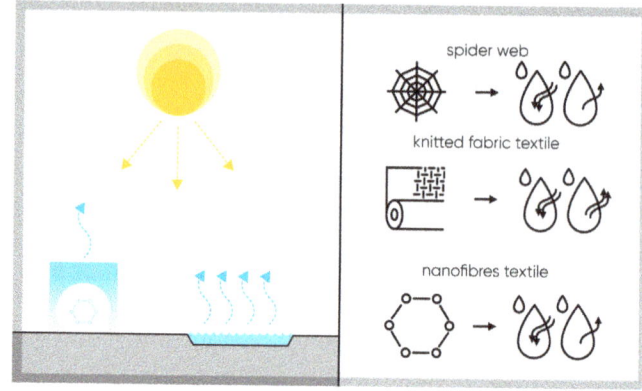

3   Conception diagram and the idea of material philosophy
    (Jan Koníček, 2021)

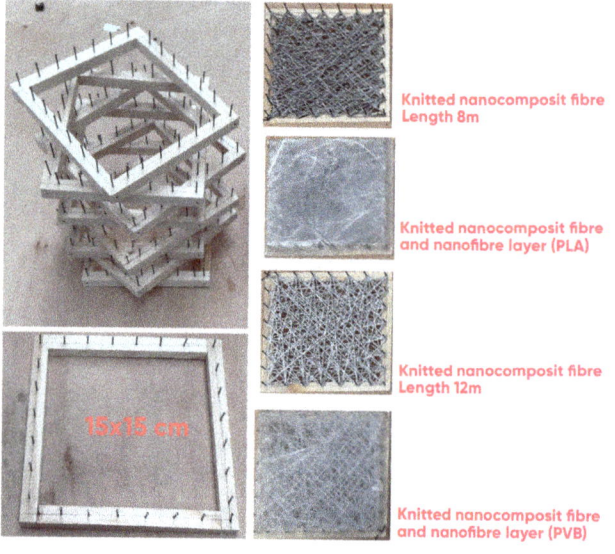

4   Testing patterns and material examples (Jan Koníček, 2021)

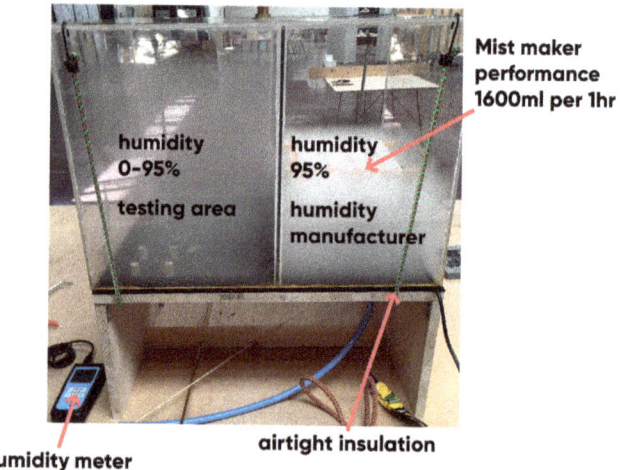

5   Humidity test box, showing the method of measuring individual samples in
    different humidity conditions (Jan Koníček, 2021)

surface area. Thanks to this, water should be retained in the nanofibrous structure for a longer time, even in sunny weather (Figure 3).

WHAT MATERIAL

The research hypothesis tests the similarity of the material properties of spider silk and nanofibers, their high specific surface area, and subsequent water adhesion. We selected a portfolio of polymers to produce nanofibrous according to the following criteria: a) water resistance, b) biodegradability, c) availability, and d) production process. We designed a simple experiment to confirm the hypotheses about the sorption properties. On the fibers of the same length, we applied another layer of nanotextile ("spider silk") (Figure 4). Specifically, these were (PVB) polyvinyl butyral,

(PCL) polycaprolactone, and (PLA) polylactic acid. We built a "fog machine" to control and distribute the fog to the samples. Finally, the weight difference over time in individual samples from the Fog machine was compared (Korczak and Urszula 2021) (Figure 5). Thanks to this method, we obtained data on the sorption and adsorption properties of individual materials within the project and confirmed the possible usable potential of nano textile structures (Figure 6).

HOW TO BUILD

The main question of the second research part was the production of nanofibrous architectural structures. We used a "nanofiber centrifuge" during the production process (Figure 7). The nanofiber centrifuge produces nanofibers

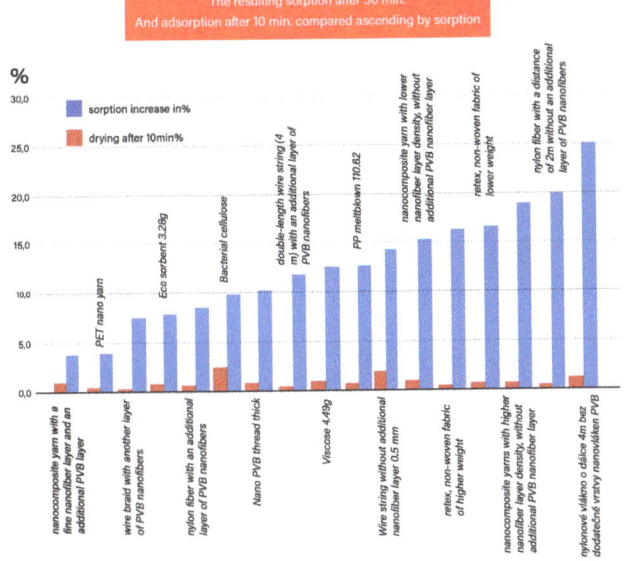

The resulting sorption after 30 min. And adsorption after 10 min. compared ascending by sorption

■ sorption increase in%
■ drying after 10min%

6 Data resulting from measuring the retention properties of a selected portfolio of materials (Klára Masnicová, 2021)

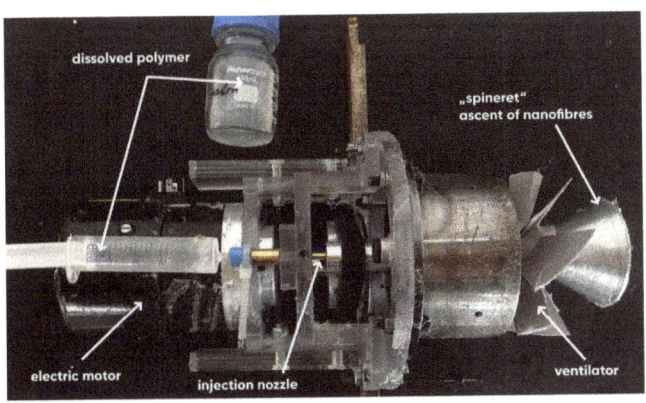

8 Nanofiber centrifuge, the mechanical part showing motor and sub-mechanical components (Jan Koníček, 2021)

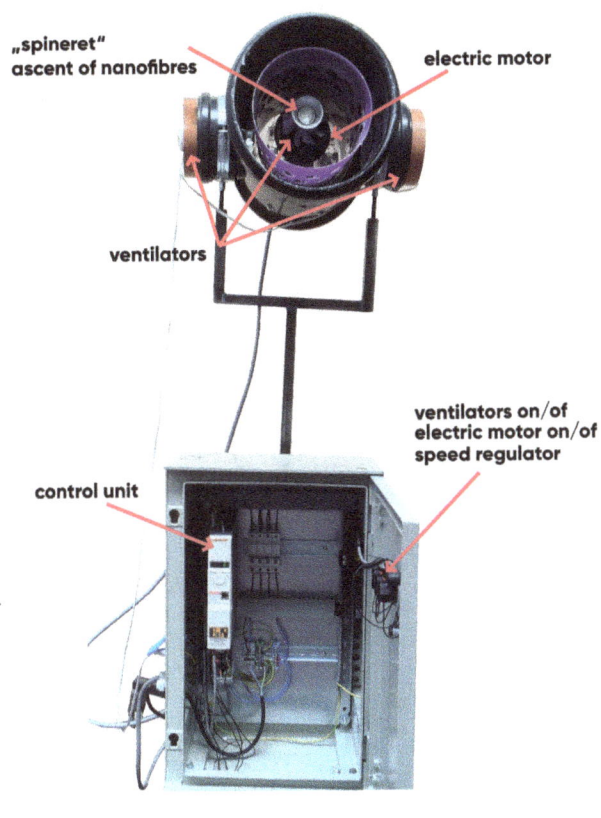

7 Nanofibre centrifuge, function prototype (Jan Koníček, 2021)

by centrifugal force, which has several advantages over electrostatic centrifugation (Williams et al. 2018). Our team made the nanofiber centrifuge. The advantages of the nanofibrous centrifuge mainly consist of the fast, easy, and safe application of nanofibrous structures. A nanofiber centrifuge can produce nanofibers from dissolved polymers in the interior and also in the exterior (Figure 8). A nanofiber centrifuge simultaneously produces fibers in micro-scale and nano-scale dimensions (Figure 9). In the future, we anticipate using robotic systems using computers with precise, controlled mobility (Gramazio et al. 2014).

RESULTING OBJECT

We made a PLA pavilion using the centrifuge made of polylactic acid (PLA) nanofibers (Figure 10). The PLA polymer

pavilion is 100% biodegradable (Gross 2002). The construction of the pavilion uses a hyperboloid shape similar to cooling towers, where water is supposed to condense based on a physical phenomenon called the chimney effect. The project describes the design and construction process of nanofibrous architectural structures and the application of nanofibers using centrifugal force (Figure 11). At the time of publication, the pavilion is undergoing testing, and the exact results are not available. We expect to complete the tests and publish the results in subsequent publications.

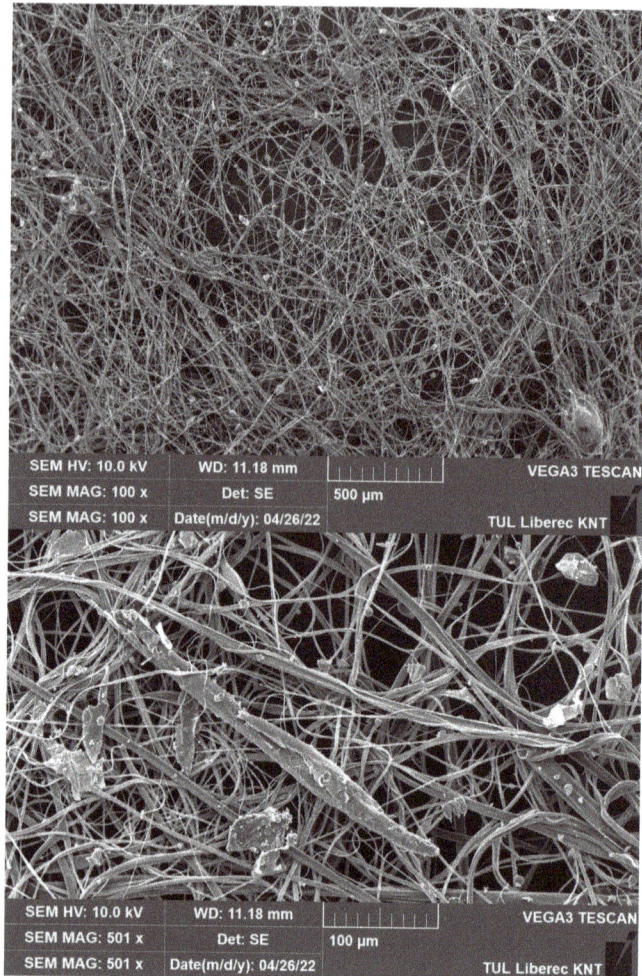

9 SEM micrography image analysis of our pavilion nanofiber structure, made by nanofibre centrifuge; (PLA) Polylactic acid polymer (Klára Masnicová, 2022)

10 Nanofiber's pavilion for catching water from air humidity (prototype) (Jan Koníček, 2021)

ACKNOWLEDGMENTS

This project results from the interdisciplinary cooperation of individual operations of the Technical University in Liberec. Thanks to all colleagues at the Department of Nonwoven Textiles and Nanofiber Materials, the Department of Textile and Single-Purpose Machines, and the Department of Architecture. Personal thanks to Jan Valter for his helpfulness and invaluable help, Jaroslava Frajová for her support and help in individual activities in the project. This research was supported by the SGS project Reg. No. SGS-2021-2043, provided by the Ministry of Education, Youth and Sports (CZE) in 2021.

REFERENCES

Finsterwalder, Rudolf. 2015. *Form Follows Nature*, 1st ed. City: Birkhäuser.

Gramazio, F., M. Kohler, and J. Willmann. 2014. *The Robotic Touch*. City: Park Books.

Gross, Richard A., and Bhanu Kalra. 2002. "Biodegradable Polymers for the Environment." *Science* 297 (5582): 803–807.

Knapczyk-Korczak, Joanna & Urszula Stachewicz. 2021. "Biomimicking spider webs for effective fog water harvesting with electrospun polymer fibers." *Nanoscale* 13: 16034-16051

Olivier, Jana. 2004. "Fog Harvesting: An Alternative Source of Water Supply on the West Coast of South Africa." *GeoJournal* 61 (2): 203–14.

Riechert, Susan E., Rosemarie Roeloffs, and Arthur C. Echternacht. 1986. "The Ecology of the Cooperative Spider Agelena Consociata in Equatorial Africa (Araneae, Agelenidae)." *The Journal of Arachnology* 14 (2): 175–91.

Williams, Gareth R., B.T. Raimi-Abraham, and C.J. Luo. 2018. "Moving from the Bench to the Clinic." In *Nanofibres in Drug Delivery*. City: UCL Press. 187–203.

11 Nanofibre application of nanofibre pavilion layer, by using Nanofibre centrifuge (Anna Machatová, 2021)

Jan Koníček is an architect, designer, and PhD student at FUA TUL. In 2020, he studied architecture at the Faculty of Art and Architecture at TUL, where he completed design studios under the guidance of doc. Ing. arch. Miloš Florián, PhD, prof. ing.arch., acad.arch. Jiří Suchomel, and Ing. Arch. Mag. by arch. Saman Saffarian. Currently, under supervisor Miloš Florián, he studies nanofiber structures and non-woven textiles in architecture at FUA TUL.

Miloš Florián worked at the Institute of Design under the guidance of prof. A. Navrátil, CSc starting in 1995. From 2002-2003, he worked at the Institute of History of Architecture and Art under the guidance of prof. P. Ulrich, CSc. Since then, he has been working at the Institute of Civil Engineering under the leadership of prof. M. Pavlík, CSc. His main areas of professional interest are glass as a structural material, intelligent building envelopes, free-form construction, and nanotechnology. He participates in scientific and research activities of ČVUT. Since October 2004, he has been leading the FLO(W) studio.

Pavel Pokorný is an employee of the Department of Nonwoven Textiles and Nanofibrous Materials TUL. His research focus includes design, construction, and implementation of technologies to prepare and produce nanofibers; design, construction, and implementation of laboratory equipment; and the study of the basic parameters of polymer solutions in a strong electrical field.

Klára Masnicová is a PhD student at the Department of Nonwoven Textiles and Nanofibrous Materials at TUL. For this project, she compiled a proposal for a portfolio of suitable materials with potential properties for the project, and she processed and evaluated the obtained data from a materials point of view.

Terrene 2.0

Biomaterial Systems and Shellular Structures for Augmented Earthen Construction

Liam Lasting, Mostafa Akbari, Laia Mogas-Soldevila, Masoud Akbarzadeh

INTRODUCTION

As our global annual CO_2 emissions continue to escalate, we must innovate within the construction industry to develop new sustainable material systems and construction methods that produce a minimum quantity of material waste. Currently, the construction industry is responsible for a total of 39% of our global CO_2 emissions, with concrete being a leading producer (UN Environment and International Energy Agency 2017). Therefore, we must source materials that require minimum energy and create a circular economy of construction materials. For specific design criteria, rammed earth construction offers solutions as a concrete substitute since it has significantly less embodied energy (Treloar et al. 2001). However, its typical construction process still utilizes wasteful formwork (Maniatidis and Walker 2003). Therefore, we propose to augment the principles of rammed earth by integrating renewable additives to improve the mechanical properties of the system towards richer geometry and developing a reusable waste-free formwork.

METHOD

Terrene's mixture (ChitoSand) consists of sand, short natural fibers, a chitosan-based binder, and a natural plasticizer, costing only $65.00 per cubic meter. Polyhedral Graphic Statics was deployed to construct a Schwarz

1 Photograph of 1m² footprint model interior with 4" tall scale figure

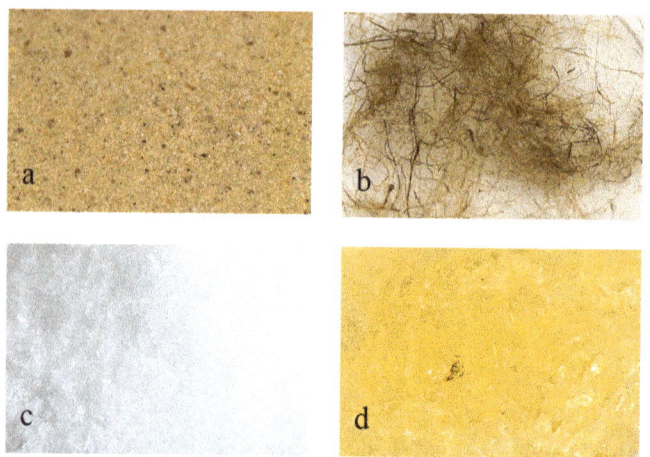

2   Material Design: (a) sand base, (b) short natural fibers, (c) natural plasti-
     cizer, (d) chitosan binder

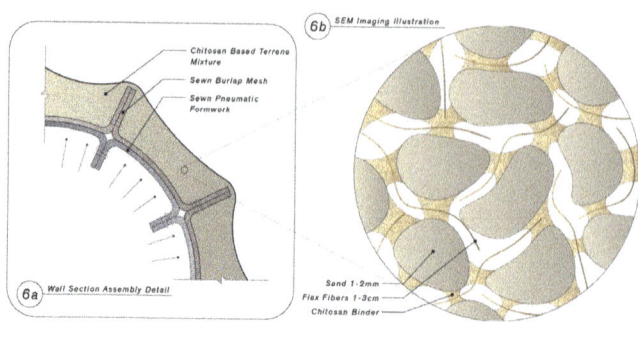

3   (a) Wall section construction/assembly detail, (b) illustration of SEM
     imaging of material composition

4   Starch-based Terrene mixture compression tested with analysis of resulting stress/strain curve and young's modulus of elasticity

P-Gyroid Hybrid Triply Periodic Minimal Surface (TPMS) geometry (Akbari et al. 2022), yielding planar surfaces to simplify the construction processes. Translated from this planar geometry, a pneumatic formwork was developed using synthetic vinyl for its durability and reusability. A corresponding burlap mesh was fit overtop the formwork. 3D-printed PLA casements framed openings and temporary tension cables fastened the formwork to the burlap. The ChitoSand was applied to the exterior surface of the burlap mesh and allowed to air dry for 24 hours before the deflation of the pneumatic formwork.

RESULTS

ChitoSand's material composition was selected to improve upon the compression-only nature of sand and provide it with a tensile capacity. ChitoSand has a reduced embodied energy due to the material system hardening over time in ambient conditions and its composition can safely be reintegrated into nature. Additionally, the Hybrid TPMS geometry was selected to show anticlastic shellular performance by taking on both tensile and compressive forces while achieving maximum strength with minimal material thickness. Initial tests using a starch-based binder resulted in a compressive strength eight times weaker than concrete (Bechthold and Weaver 2017), but still performative for minimal-material tension-compression shell structures (Lasting et al. 2022). Compressive testing of the used ChitoSand material is still underway but is expected to improve upon the starch-based results dramatically.

To help facilitate the construction of this unconventional geometry, formwork is required. Typical construction uses single-use formwork made of wood or steel (Hurd 2005). Instead, we designed a reusable pneumatic formwork to create modular units that can aggregate into a larger habitable network. By embedding its seams within the applied

Terrene 2.0 Lasting, Akbari, Mogas-Soldevila, Akbarzadeh

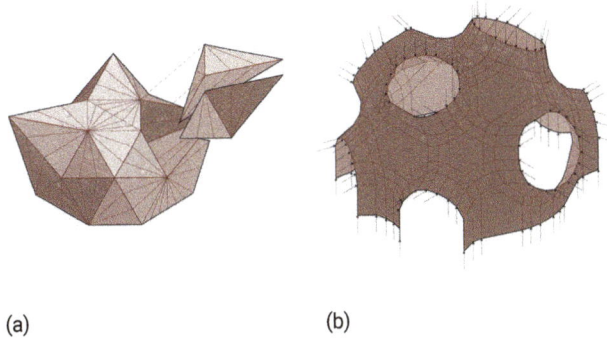

(a)   (b)

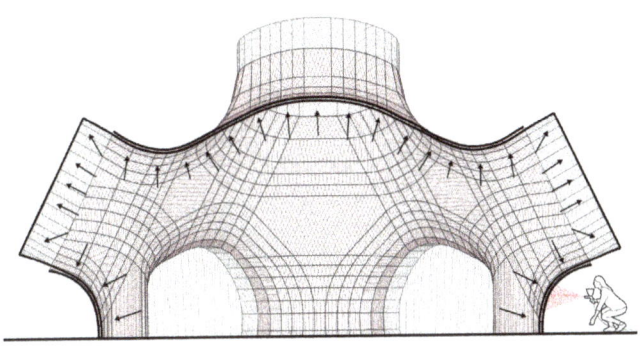

(c)

5   Polyhedral Graphic Statics Model: (a) form's polyhedral force diagram, (b) Schwarz P-Gyroid Hybrid triply periodic minimal surface (TPMS) geometry

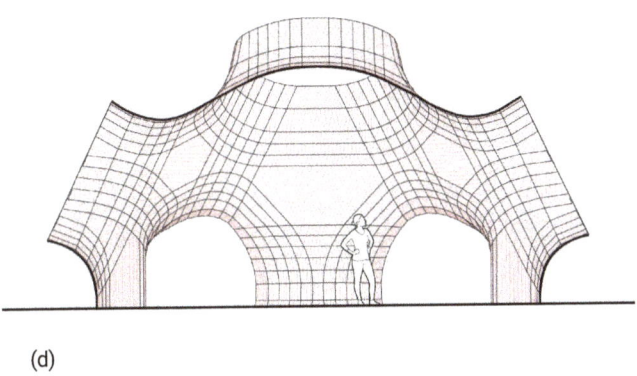

(d)

6   Completed model with 1m² footprint with 4" tall scale figure

7   (c) Pneumatic formwork inflated and ready for ChitoSand application.

    (d) Pneumatic formwork is removed once ChitoSand has dried in-situ

ChitoSand, the burlap mesh provides additional tensile strength to the material system. With all these components being abundant or naturally produced and assembled using intelligent structural design, the proposed material system and construction method are accessible, efficient, and provide an affordable and sustainable building solutions.

CONCLUSION

As we look forward to a more sustainable building industry, we must continue to explore creative material systems and construction methods that help provide alternatives to the conventional materials we have grown accustomed to. Additional research is currently underway to adapt this prototype to a habitable scale. Therefore, components such as the 3D printed casements will be substituted with wood or cork veneer and the ChitoSand material will be applied to the surface with a pneumatic shotcrete system. While this material might not be able to replace concrete altogether,

it can provide a low environmental impact substitute that can be utilized when appropriate during the design and construction process. While ChitoSand guided this particular construction project, this method can also be deployed with other materials for reduction or elimination of formwork waste.

ACKNOWLEDGMENTS

This research was supported by the University of Pennsylvania Research Foundation Grant (URF) to Dr. Laia Mogas-Soldevila. It is also partially funded by the National Science Foundation (NSF) CAREER AWARD (NSF CAREER-1944691- CMMI), and the National Science Foundation (NSF) Future Eco Manufacturing Research Grant (NSF, FMRG-CMMI 2037097) to Dr. Masoud Akbarzadeh. The authors gratefully acknowledge the use of facilities and instrumentation supported by the Materials Science and Engineering Departmental Laboratory at the University of Pennsylvania.

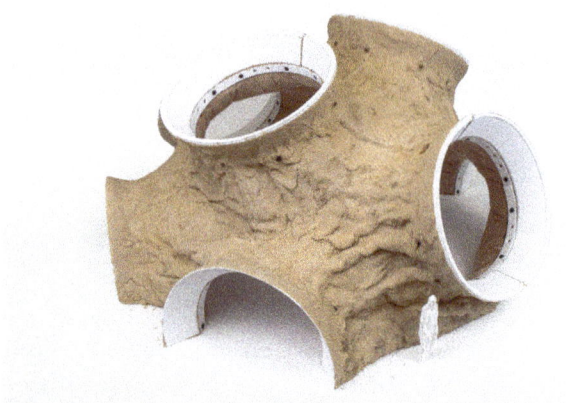

9   Completed model with 1m² footprint with 4" tall scale figure

10   Photograph of 1m² footprint model interior with 4" tall scale figure

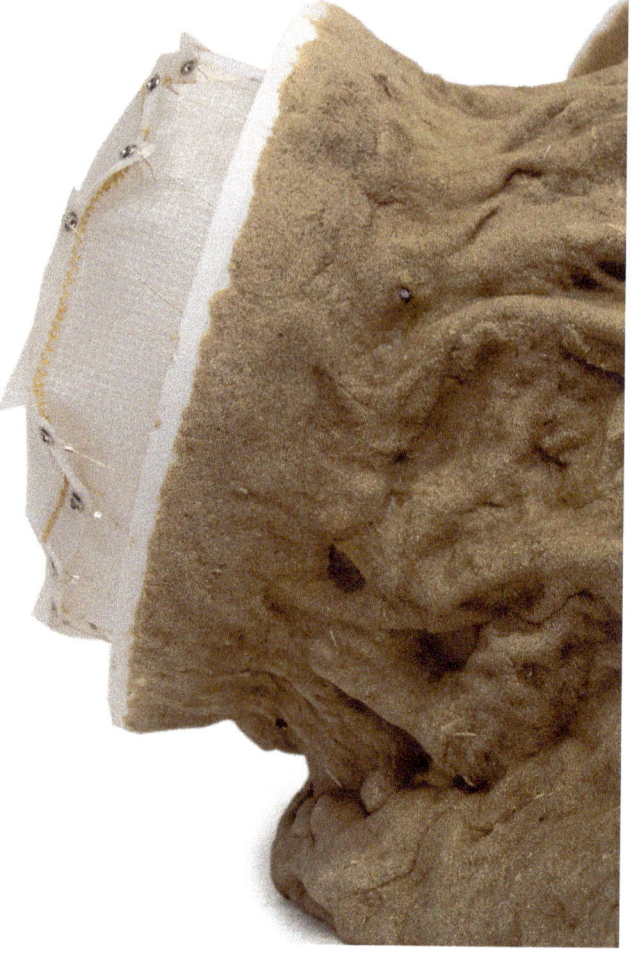

11   Inflated pneumatic formwork in place while ChitoSand is drying

REFERENCES

Akbari, Mostafa, Armin Mirabolghasemi, Mohammad Bolhassani, Abdolhamid Akbarzadeh, and Masoud Akbarzadeh. 2022. "Strut-based Cellular to Shellular Funicular Polyhedral Materials." *Advanced Functional Materials* 32 (14): 2109725.

Bechthold, M. and J.C. Weaver. 2017. "Materials science and architecture." *Nature Reviews Materials* 2 (12): 17082.

Hurd, Mary Krumboltz. 2005. *Formwork for Concrete.* Farmington Hills, IL: American Concrete Institute.

Lasting, Liam, Isabelle Lee, Laia Mogas-Soldevila, and Masoud Akbarzadeh. 2022. "Terrene 1.0: Innovative, Earth-Based Material for the Construction of Compression-Dominant Shell Structures." In *Proceedings of IASS Annual Symposia, International Association for Shell and Spatial Structures (IASS)*, vol. 2022. 1–12.

Maniatidis, Vasilios, and Peter Walker. 2003. "A Review of Rammed Earth Construction." Paper presented at DTI Innovation Project Report, Natural Building Technology Group. Bath, UK: University of Bath. 12.

Treloar, Graham J., Ceridwen Owen, and Roger Fay. 2001. "Environmental assessment of rammed earth construction systems." *Structural Survey* 19 (2): 99-106.

UN Environment and International Energy Agency. 2017. "Towards a zero-emission, efficient, and resilient buildings and construction sector; Global Status Report 2017."

IMAGE CREDITS

All drawings and images by the authors.

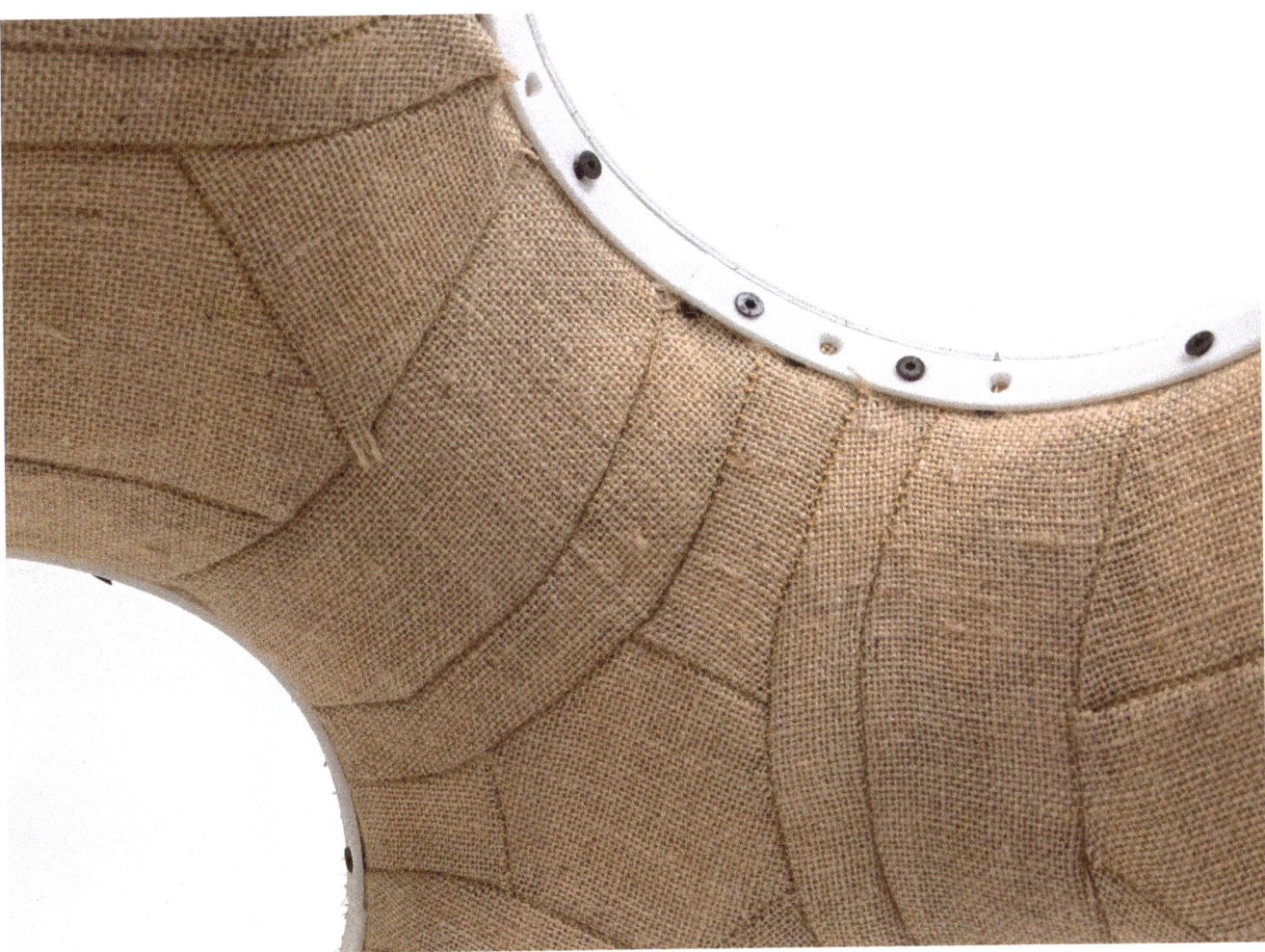

12 Interior connection and texture detailing photograph

Liam Lasting is an architectural designer and recent Masters of Architecture graduate from the University of Pennsylvania's Stuart Weitzman School of Design. Originally from Southern California, he holds a bachelor's degree in Architectural Studies and a minor in Urban and Regional Studies from the University of California, Los Angeles. His graduate architectural thesis has initiated a path of research through a position at the Polyhedral Structures Laboratory and DumoLab where he can explore the intersections of computational structural design and sustainable biomaterial formations.

Mostafa Akbari is a computational designer and researcher with a background in Architectural design and computation. He is currently a PhD student at the Polyhedral Structures Laboratory, University of Pennsylvania. He has graduated from Master of Advanced Architectural Design from Stuart Weitzmann School of Design for which he received the highest merit-based scholarship based on excellent qualifications. He holds a Bachelor of Architecture at the University of Tehran and a Master of Architecture at the University of Shahid Beheshti.

Laia Mogas-Soldevila is an Assistant Professor of Graduate Architecture and Director of DumoLab Research at the Stuart Weitzman School of Design, University of Pennsylvania. Her broader pedagogy supports novel theory and applied methods understanding materials and materialization in architecture. Laia holds an interdisciplinary doctorate bridging materials science, biomedical engineering, and design from Tufts University School of Engineering, two master's degrees from the Massachusetts Institute of Technology, and is a licensed architect with a minor in Fine Arts by the Polytechnic University of Catalonia School of Architecture.

Masoud Akbarzadeh is a designer with academic background and experience in architectural design, computation, and structural engineering. He is an Assistant Professor of Architecture focusing on Structures and Advanced Technologies and the Polyhedral Structures Laboratory (PSL) director. He holds a D.Sc. from the Institute of Technology in Architecture, ETH Zurich where was a research assistant in the Block Research Group. In addition, he has two degrees from MIT: a Master of Science in Architecture Studies (Computation) and a MArch, the thesis for which earned him the renowned SOM award.

Controlled Buckling

Ultra-Thin Folded Paper Formwork

Joseph Choma, Ena Lloret-Fritschi, Fabio Scotto, Anna Szabo, Robert Flatt

In this research, material-lean customizable formwork for concrete is developed by stiffening such molds through the use of folded paper. For this, the research capitalizes on the large body of work that has gone into paper folding—such as mathematical geometry and mechanical behavior of hinges—to design mechanically robust folded structures. This leads to molds that can withstand hydrostatic pressure, facilitating the integration of digital concrete processes. By using paper, a formwork is produced that can be peeled-off, recycled in established material streams, and leave a surface finish matching the highest expectations for architectural concrete.

How can folding advance concrete casting? When approximately 40% of the carbon dioxide emissions in the world are associated with the built environment, there is an ethical obligation to transform how we design and build more sustainably (International Energy Agency 2019). Additionally, concrete is by far the most used building material in the world (Gagg 2014). One research approach would be to develop new means and methods to build with natural materials that have the potential to replace concrete, such as large scale timber structures (Foster and Ramage 2020). Our research accepts concrete as a fundamental building material (even if it is unsustainable) and questions how its casting method can be improved. Although shell structures (Calladine 1983) are

PRODUCTION NOTES

Funding: SPARK SNSF Grant

NCCR Digital Fabrication Researcher in Residence

Location: Zürich, Switzerland

Date: 2019–2021

1   Detail of a folded paper formwork that incorporates nested curved creases

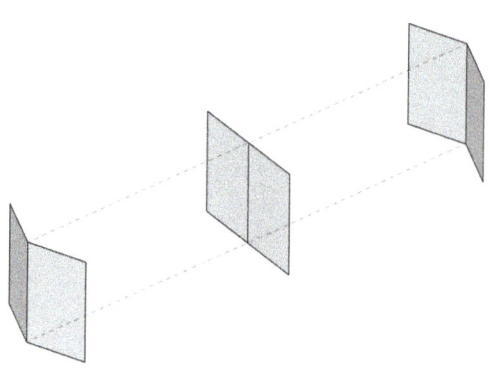

2 Diagram of a simple hinge, where the folded geometry is always unstable

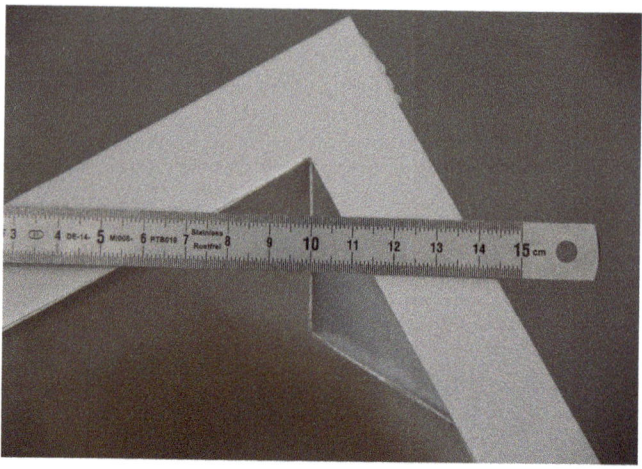

4 Formwork from the top view; the triangular section is a nested curved crease

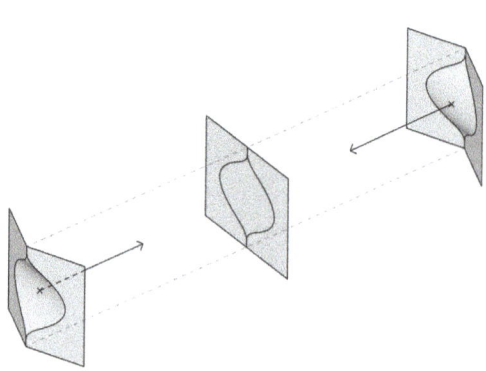
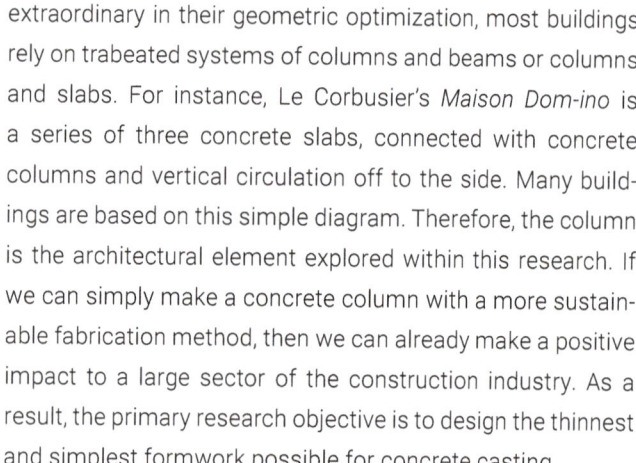

3 Diagram of a bistable structure, where the geometry locks at a particular angle

5 Formwork is able to resist the transient hydrostatic pressure of the concrete

extraordinary in their geometric optimization, most buildings rely on trabeated systems of columns and beams or columns and slabs. For instance, Le Corbusier's *Maison Dom-ino* is a series of three concrete slabs, connected with concrete columns and vertical circulation off to the side. Many buildings are based on this simple diagram. Therefore, the column is the architectural element explored within this research. If we can simply make a concrete column with a more sustainable fabrication method, then we can already make a positive impact to a large sector of the construction industry. As a result, the primary research objective is to design the thinnest and simplest formwork possible for concrete casting.

Although it is not fair to directly compare mass-produced concrete elements with bespoke elements, it is beneficial as a means to understand a larger spectrum of the industry. Typically, mass-produced concrete elements use heavy duty steel formwork that produce the same part over a thousand times. These types of formwork are usually extremely heavy

and difficult to move and thus are usually used in pre-fabricated environments (off-site). Timber and plywood formwork are commonly used for on-site casting. Much wooden formwork can be reused up to five times before being discarded (Halvorsen 1993). On the other hand, bespoke formwork are usually extremely expensive and time intensive to fabricate and are usually used only once before being discarded to landfills. For example, the Rolex Learning Center by SANAA was fabricated using custom wooden formwork that were all discarded after one time use (Weilandt et al. 2009). Formwork for standard concrete elements can makeup approximately 50% of the budget for that architectural element (Lab 2007). This usually equates to about 10% of an entire building's construction budget (Hanna 1998). For nonstandard concrete elements the formwork can be as much as 75% of the budget for that architectural element (Schipper and Grünewald 2014). This research looks to use inexpensive ultra-thin folded paper formwork (1/64 inch or 0.4 mm) that is used once but can be

Controlled Buckling Choma, Lloret-Fritschi, Scotto, Szabo, Flatt

6 Peeling the paper formwork off like a candy wrapper

7 The 1-meter tall cast concrete column

fully recycled after its use. In theory, the folded system could also allow for more customized bespoke variations without significant additional costs.

UNSTABLE TO BISTABLE

The column formwork design began by challenging a simple hinge or linear crease. Typically, single creases behave like a hinge, such that the fold is never fixed at any particular angle. For example, you can open and close a book at any angle. However, by incorporating curved creases that "kiss" and mirror along the vertical axis, it is possible to create bistable structures. By carefully calibrating the arc of the curved creases it is possible to control the maximum angle that it can be folded. In other words, a stouter curved crease will be able to fold at a shallower angle than one with an elongated curved crease. As the curved creases are folded, the inner "cell" defined by the creases will "pop" inward. This is similar to controlled buckling (Lee 2019) and a curved tape

spring (Seffen et al. 2000), transforming the simple hinge at a corner to a bistable structure. This behavior must be specifically designed into the crease pattern such that the equilibrium shape is achieved that will in turn help to accommodate transient hydrostatic pressure, which is reduced but not fully eliminated by the set-on-demand concrete process. This initial design approach was tested with a handful of fabrication-based experiments, yielding promising results.

Nested curved creases were also added to the design. Nested creases—arrayed curved creases that alternate between mountain and valley folds—provide increased surface area around the formwork. This feature can contribute to the structural performance of the ultra-thin formwork; however, it needs to be carefully calibrated. For certain designs and geometries, the incorporation of a valley fold can make the formwork weaker. For example, for a very elongated curved crease, it will be easy to pop the valley fold outward; and for a very stout curved crease, it will be almost impossible to fold

8   Casting using an ultra-thin folded paper formwork

9   The tapering concrete column with a height of 2.5 meters

the initial valley fold at all. Overall, the nested creases demonstrated an increased rigidity of the folded formwork. In the next fabrication-based experiment, the formwork was made of a wax coated paper with a material thickness of 1/64 inch or 0.4 mm.

SCALING UP

Next, our research started to tackle a full-scale prototype that combined the efficiency of curved creased paper formwork with the controlled set-on-demand concrete casting process. The prototype was a 2.5 meter tall tapering column produced with industry-grade concrete. This mixture had aggregate sizes up to 8 mm and a compressive strength that reaches 85 MPa after 28 days. The set-on-demand casting process used a combination of chemical admixtures to ensure constant control over the hydration phase of the concrete. For this prototype, the casting flow rates varied from 1.5 to 3.0 liters per minute to respond to the cross-sectional changes of the column. The total casting time was approximately 100 minutes.

This prototype successfully demonstrated an alternative approach to fabricating concrete elements by combining set-on-demand casting with folded paper formwork. This study lays the basis for a systematic rethinking of the design and performance of concrete formwork, providing an innovative solution for the future of manufacturing.

REFERENCES

Calladine, C. R. 1983. *Theory of Shell Structures*. Cambridge: Cambridge University Press.

Foster, R.M., and M.H. Ramage. 2020. "Tall Timber." In *Nonconventional and Vernacular Construction Materials: Characterisation, Properties and Applications*, 2nd ed., edited by K.A. Harries, and B. Sharma. Duxford, UK: Woodhead Publishing. 467–490.

Gagg, C.R. 2014. "Cement and Concrete as an Engineering Material: An Historic Appraisal and Case Study Analysis." *Engineering Failure Analysis* 40, 114–140.

Halvorsen, GrantT. 1993. "Form Reuse." *Concrete Construction Magazine*. Online. March 1, 1993. Accessed Sept 5, 2022. https://www.concreteconstruction.net/products/concrete-construction-equipment-tools/form-reuse_o.

10 Detail photograph demonstrating the scale of a hand; in general, the column prototype had a smooth "waxy" surface finish

Hanna, Awad S. 1998. *Concrete Formwork Systems*. New York: Marcel Dekker.

International Energy Agency. 2019. *2019 Global Status Report for Buildings and Construction: Towards a Zero-Emission, Efficient and Resilient Buildings and Construction Sector.* Paris: International Energy Agency.

Lab, Robert H. 2007. "Think Formwork—Reduce Costs." *Structure Magazine* April 2007: 14–16.

Lee, T. 2019. "Elastic Energy Behaviours of Curved-Crease Origami." PhD diss., University of Queensland.

Schipper, H.R., and S. Grünewald. 2014. "Efficient Material Use Through Smart Flexible Formwork Method." In *ECO-Crete: International Symposium on Environmentally Friendly Concrete.*

Seffen, K.A., Z. You, and S. Pellegrino. 2000. "Folding and Deployment of Curved Tape Springs." *International Journal of Mechanical Sciences* 42 (10): 2055–2073.

Weilandt, A., M. Grohmann, K. Bollinger, and M. Wagner. 2009. "Rolex Learning Center in Lausanne: From Conceptual Design to Execution." In *Proceedings of the IASS Symposium: Evolution and Trends in Design, Analysis and Construction of Shell and Spatial Structures.* 640–653.

IMAGE CREDITS

All photographs, drawings, and images by the authors.

Joseph Choma is Director of the School of Architecture and Professor of Architecture at Florida Atlantic University.

Ena Lloret-Fritschi is Assistant Professor at the Academy of Architecture in Mendrisio at Università della Svizzera italiana (AAM-USI).

Fabio Scotto is XR Integrator at Pixelmolkerei AG in Chur, Switzerland. He completed graduate studies at ETH Zürich and University of Edinburgh.

Anna Szabo is Junior Scientist at Sika. She completed her PhD and postdoctoral research at ETH Zürich.

Robert Flatt is Professor of Physical Chemistry of Building Materials at Institute of Building Materials, ETH Zürich.

Orbital

Computational Garden Folly

Jason Kelly Johnson, Nataly Gattegno

Orbital is a computational garden folly, exploring geometric and material exuberance, digital design and fabrication. It evokes not only organic forms found in nature, but also giant robots and futuristic space vehicles. The structure is composed of three coiled legs that spiral towards the sky. The exterior double-curved surfaces are defined by stainless steel origami skins, while the interior space is wrapped by a vortex of colorful tactile aluminum shingles. *Orbital*'s dynamic form evokes an era of rapid change and uncertainty, while also inspiring curiosity and playful interaction.

Orbital was commissioned as a "percent-for-art" public artwork. It was awarded through an invited two-phase competition. The sculpture is sited within a public space adjacent to Uber's new global headquarters in San Francisco's Mission Bay. The site was originally a coastal marsh. While the sculpture is lightweight and only 33 feet tall, the landfill subsurface required the artwork to be supported on three approximately 90-foot deep piles.

Our design-build studio FUTUREFORMS produced all of the design and construction documents, shop drawings, and assembly drawings for the project. The stainless steel superstructure and skin supports were fabricated in Olson Steel's nearby welding shop. The sculpture's stainless steel skins were cut and bent by local vendors, then assembled by our in-house team. The colored anodized aluminum shingles were produced in Neal Feay's Santa Barbara CNC shop, then assembled and installed by our team. Our team worked alongside a team of professional riggers and installers to complete the installation of the project in under two months during the pandemic.

CONCEPT

Orbital spirals into the sky, formed by a curved trajectory around the center of the sculpture, to create a intimate place beneath it. The geometry is initially repetitive and rational, and then unexpectedly transforms into coiled orbitals of steel and color. The exterior and interior skins of the structure are geometrically similar, but materially distinct tessellated and shingled modules. The highly reflective exterior is contrasted by a tactile and colorful interior (Figure 2).

Orbital explores ways of bringing together these two contrasting conditions by creating an artwork that can be experienced in two distinct ways. On the outside a textured, tessellated, high-definition surface articulates a towering structure that anchors the site, is visible from afar, and attracts and provides visual identity. On the inside a color

PRODUCTION NOTES

Artists:	Jason Kelly Johnson & Nataly Gattegno (FUTUREFORMS)
Credits:	www.futureforms.us/orbital
Engineer:	Arup
Client:	Uber Technologies
Location:	Mission Bay, San Francisco, California
Date:	2021

1   View of the *Orbital* sculpture and the neighboring public space

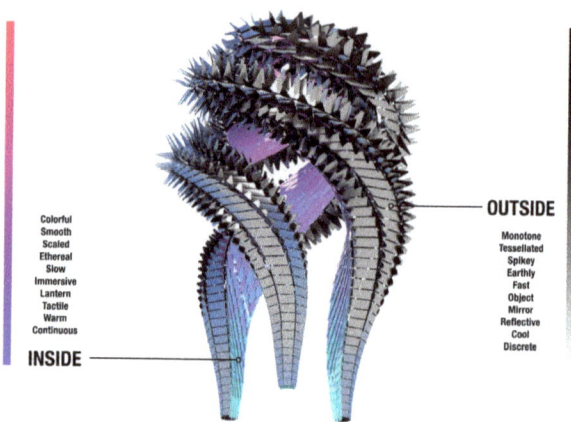

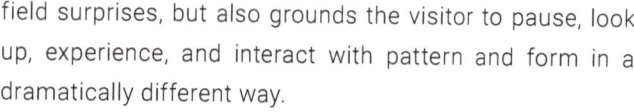

2 Concept diagram illustrating the qualities of inside versus outside

3 Exploded diagram of the coiled leg assembly

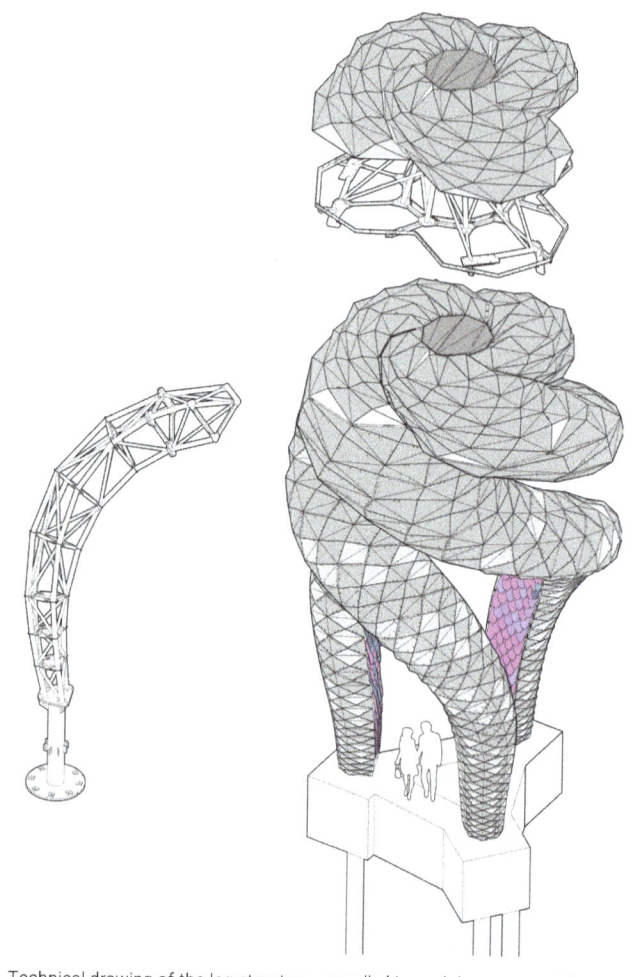

4 Technical drawing of the leg structure, overall skin, and dome assemblies

field surprises, but also grounds the visitor to pause, look up, experience, and interact with pattern and form in a dramatically different way.

This binary "inside outside" experience of the artwork extends into its daytime and nighttime presence. During the day, *Orbital*'s exterior takes over with hints of the colorful interior that peek through and which are only revealed to those who step under or walk through and look up. At night, the interior colored surface is illuminated with in-ground flood lights. This gives it an internal glow and makes *Orbital* perform as a very different object, glowing and appearing like a lantern in a garden. Slits of light escape the interior to create a very different reading of the geometry as it sweeps upwards.

Orbital engages directly with the site and the public. Highly reflective, it creates an ever-changing fractal image of the surroundings on its surface. *Orbital* also encapsulates a second way of interacting and experiencing the artwork

when one steps under it and is immersed in a colorful space shaped by the three twisting forms. The space is smaller in scale, to create a much more intimate experience for fewer people. This inner space culminates in an oculus that is open to sky. In the summer months a shaft of sunlight dances around the ground and reflects rays on the inner space. These fast moving rays create unexpected auroras of colored light and specular reflections.

COMPUTATIONAL DESIGN

The centerline guiding each of the structure's coiled legs consists of a parametric helical curve. The endpoint of this curve converges on a circle just below the base of the upper dome (Figure 4). Each identical curve is copied and rotated 120 degrees to define the other two centerlines. Each curve rail then defines the vertical sweep of the sectional geometries. The variable sections are straight on the inner shingle side and curved on the outer stainless panel side (Figure 3). The three panelized coiled legs converge to form a faceted

5 Photo of the stainless steel structural tube assemblies being erected on site

7 View of the coiled leg assembly: aluminum shingles and stainless steel pans

6 Installation view of the prefabricated outer skin assemblies

8 One of three dome structural skin assemblies being hoisted into place

dome on the outside, and the inner aluminum shingles converge to form a nine-sided mirrored oculus. Rhino and Grasshopper were used as the primary design, simulation, and documentation tools throughout the process.

PROTOTYPING

Throughout the process of designing *Orbital*, our team physically prototyped the global geometries, details, skins, and assemblies at various scales. Full scale mock-ups were constructed to explore and verify tolerances, ease of assembly, installation and future maintenance.

DIGITAL CRAFT

A range of digital fabrication tools were used throughout the process. All of the faceted outer skins were cut using a fiber laser and bent using a CNC press break. The colored anodized aluminum shingles were each created using a HAAS 5-axis CNC mill station. The base stainless steel helical tube structure was created using a CNC tube cutter that allowed

each complex and unique part to be cut with precision. After the parts were welded together, the stainless steel panel frames were bolted to the structure in the shop.

INSTALLATION

After the structure was transported and bolted together on site, the skins were installed in large, prefabricated chunks. While the lower portion of each leg's skins is supported by a traditional webbed truss, the dome is a structural skin that was lifted into place as three large prefabricated units (Figure 8). The exterior perforated skin consists of hundreds of bent panels, while the interior skin was created using hundreds of overlapping colored anodized shingles. Each shingle was installed from the bottom up by the author's studio.

ACKNOWLEDGMENTS

Lead Artists: Jason Kelly Johnson & Nataly Gattegno (FUTUREFORMS)

Artist Team: Jason Kelly Johnson and Nataly Gattegno with Carlos Sabogal, Brian McKinney, Clayton Williams, Natalie Abbott, Chris Leo,

9   View of *Orbital*'s outer skin taken from the terrace above the public garden

10   Photo of *Orbital*'s dome and oculus taken from a neighboring parking garage

Valerie Tse; assisted by Lee Marom, Ki Schmidt, and Sam Higgwe

Structural Engineer: Arup (Nick Sherrow-Groves, Lead), Felix Weber

Fabrication Team: Futureforms (San Francisco, CA), Olson Steel (San Leandro, CA), Seaport Stainless (Richmond, CA), Standard Sheet Metal; Skin installers: Pacific Erectors, Sheedy Crane Co.; Aluminum shingles: Neal Feay (Santa Barbara).

Landscape Architect: Surface Design
Building Architect: SHOP Architects
Art Consultant: Dorka Keehn

IMAGE CREDITS

Photos 1, 9, 10, 11: ©Matthew Millman Photography, 2021
All other drawings and images by the authors

Jason Kelly Johnson (born 1973, Canada) is lead artist and founding design partner of FUTUREFORMS. He brings an expertise in computational design and advanced digital fabrication through the lens of critical art production and interactive technologies. Jason was born and raised in Canada. He received a Bachelor of Science from the University of Virginia, and a Masters of Architecture from Princeton University. Johnson is currently a tenured Professor at the California College of the Arts in San Francisco, where he serves as the co-director of CCA's Digital Craft Lab. He formerly served as the President of ACADIA (the Association for Computer Aided Design in Architecture) and a member of its Board of Directors for many years.

Nataly Gattegno (born 1977, Greece) is an artist and founding managing partner of FUTUREFORMS. She brings an expertise in design research and urban speculation, through the lens of art and urban design. Nataly was born and raised in Athens, Greece. She received a MA from Cambridge University, St. John's College, UK, and a Masters of Architecture from Princeton University. Gattegno is currently a Professor at the California College of the Arts in San Francisco, CA.

FUTUREFORMS is an award-winning art and design studio based in San Francisco. Founders Jason Kelly Johnson and Nataly Gattegno have collaborated on a range of projects exploring the intersections of art and design with public space for over 20 years. Recent public art projects have included sculptural shade canopies, art pavilions, fine art objects, furniture and lighting, as well as large scale urban art installations and art master plan consulting. www.futureforms.us @futureformslab

11   View from below of *Orbital*'s three coiled legs and inner colored anodized aluminum shingles

ROCKING CRADLE

Reconstituting Geology on a Damaged Earth

Dana Cupkova, Matthew Huber, Edith Abeyta

The *Rocking Cradle* is an interactive installation comprised of multiple binder-jet-printed vessels integrated into a public tree nursery. Informed by hydrological and geological processes, the *Rocking Cradle* is conceived as a literal and figurative device for fostering environmental stewardship. Situating the tree nursery as a community space, these sand-printed vessels hybridize the typology of urban furniture with the behaviors of water flow, water collection, bird bathing, and growing native plants. Located in a polluted post-industrial landscape, the installation is conceived as a new form of ecological infrastructure that fosters stewardship through the instigation of urban play (Cupkova and Huber 2021). Co-authored through a series of environmental justice workshops, the semi-porous, stone-like surfaces carry embedded messages from local youth, thus becoming vehicles for local environmental consciousness intertwined with communal discourse.

The project engages the former steel mill site and its pollution patterns by drawing new awareness to its conflicted histories. Each vessel serves as a symbolic substrate that is formed from anthropogenic and earthen waste to create both a protective barrier and remediated ground for nurturing new plant life. Carrying forward technologies of water flow simulation (Cupkova, Azel, and Mondor 2015), the articulated surfaces build up a visual intuition for the ways in which landscapes behave entangled with cryptic text/plant graffiti to inspire new forms of empathy. The objects, through their geometric figuration, begin to care for the landscape they sit in. They cultivate new ecological

PRODUCTION NOTES

Design Team:	Epiphyte Lab, SoA CMU, Arts Excursion Unlimited
Client:	Center of Life
Status:	Built
Site Area:	40,000 sq. ft.
Location:	Hazelwood Green, Pittsburgh, Pennsylvania
Date:	2021

1   Tree nursery installation view of the Planter Rocking Cradle with carnivorous swamp plant at Hazelwood Green (©Massery Photography, 2022)

2 Sand-printed surface texture (©Epiphyte Lab 2020)

3 Nature drawing workshop (©Arts Excursions Unlimited 2020)

4 Calcification and new habitat growth due to surface rugosity (©Epiphyte Lab 2022)

5 Rock, mud and rainwater play (©Epiphyte Lab 2022)

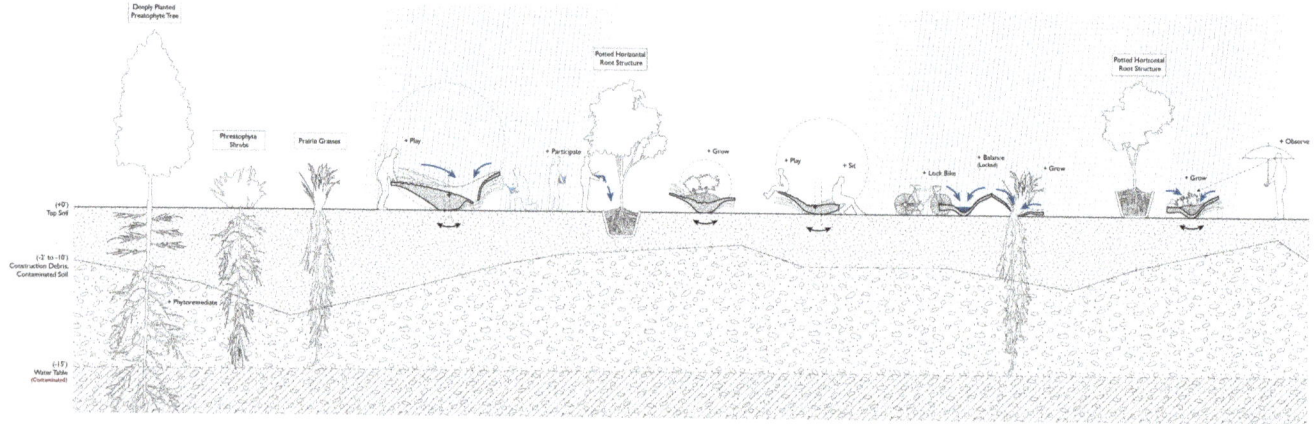
6 Environmental stewardship concept diagram: *Rocking Cradle* variations exploring Human-Land-Seedling-Hydrology-Community engagement (©Epiphyte Lab 2020)

habitats and human interactions, thus opening up new modes of play and ecological intimacy.

The *Rocking Cradle* is also a prototype rooted in ongoing material-technology research that proposes a novel cradle-to-cradle design process for architectural components to be manufactured directly from local construction waste and earthen materials. The body of the cradle combines shaping strategies for volume (Craveiro et al. 2017) and surface figuration with complex patterning (Dunn and Halpin 2009) derived from balancing behaviors (Clifford et al. 2019) to enable the rocker to become a structural component, as well as a substrate for ecological processes.

Such shaping strategies consider the life-cycle of construction from a cradle-to-cradle perspective (Faludi et al. 2019). Cementitious materials such as concrete, because of the volume used, are some of the greatest contributors to the global increase of CO_2 levels. Two interconnected strategies are employed here to subvert this status quo:

reducing construction waste by using the material in binder-jet manufacturing, and reducing the overall volume of the material used through material-specific shape-sensitive component design (ExOne 2022). By advancing the additive manufacturing of earthen and non-cementitious materials, CO_2 production can be significantly reduced. This approach shifts issues of advanced manufacturing into a framework of ecological design. Shape-factor plays a central role in the formation of the cradle, tuning volume distribution within the mass, and combining structural and ecological factors with a composite of granular waste materials.

Rocking Cradle aligns structural and ecological potential (Cupkova and Clifford 2018) with a desire to integrate landscape awareness, its history, and presence directly into the architectural form, behavior, and experience. Surface articulation enabled by binder-jet-ting technology provides an opportunity for traces of text—a voice that typically would only be transient—to be embedded into and carried forward more permanently within the object

7   Text graffiti workshop at the Center of Life
    (©Arts Excursions Unlimited 2021)

8   Joy of water play on a hot day at the Bird Bath Rocking Cradle
    (©Massery Photography 2022)

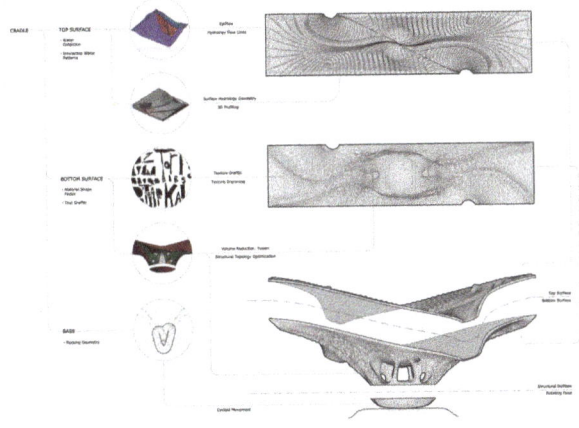

9   Design workflow diagram of computational shaping integrates water
    pathways, text graffiti, and balancing, while using minimal material mass
    (©Epiphyte Lab 2022)

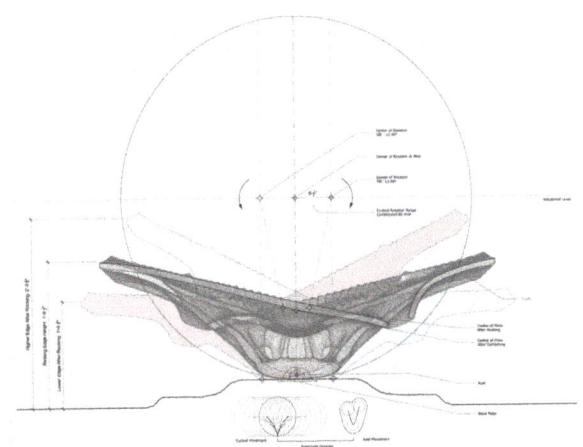

10 Diagram of mass balancing strategy using cycloid movement
    (©Epiphyte Lab 2022)

itself. The layered operation of shaping the vessels also allows for a collective adaptation of the form and carving of messages to communicate growth. The *Rocking Cradles* care for and support human-land-seedling-hydrology-community engagement while connecting the past with the future. As extreme climate actions intensify, questions of public space will increasingly elide with those of planetary ecology, and the boundary-challenging expansiveness of ecological flow. Arising from entanglements of landscape abuse and the impact on humans and non-humans alike, the *Rocking Cradle* actively considers revitalization and emergence of new forms of urban gardening. As we grow food and oxygenating plants in this contaminated soil, we precipitate the landscape to body-contamination pathways. *Rocking Cradle* intends to offset the current cycle while using its body as a device for future land stewardship.

Urban play also offers an immense opportunity as a vector of curiosity and provocation of intimacy and empathy between humans, objects, and landscapes. Arising anew from an ossified post-industrial landscape, the *Rocking Cradle* intertwines human interaction and the ecological imagination.

ACKNOWLEDGMENTS

The *Rocking Cradle* project has been conceived as a part of the funded research "CRuMBLE: Construction Rubble Manufacturing for Building Life-cycle and Environment," a research-manufacturing framework that proposes a novel cradle-to-cradle design process for ecological architecture manufactured directly out of construction waste and earthen materials by integrating recycled construction waste as a powder aggregate mixture to create new pathways for direct non-toxic chemical activation with water-based binders in binder-jet printing. PIs: Dana Cupkova, Josh Bard, CoPI: Robert Heard; Industry Partners: ExOne, Michael Brothers Hauling.

This project has been supported by The Manufacturing Futures Institute at Carnegie Mellon University (CMU), The Manufacturing PA Innovation Program, ExOne Company, CMU School of Architecture, The Pittsburgh Parks Conservancy, Center of Life Hazelwood, The Fund for Research and Creativity College of Fine Arts CMU, U3 Advisors, Rivers of Steel's Mini-Grant Program, Arts Excursion Unlimited, and EPIHYTE Lab.

SoA CMU Team: Design and Research Lead: Dana Cupkova; Design

11 Arts Excursions Unlimited team with Planting Rocking Cradle: Edith Abeyta with two high school students,Tayshaun Watkins and Samuara Green, enrolled in the Start on Success Pittsburgh program (Photo ©Lake Lewis)

Development Lead: Matter Huber, Environmental and Community based Art Direction: Edith Abeyta; Design Development and Production Team: Marantha Dawkins, Kirman Hanson, Gil Jang, Ryu Kondrup, Longney Luk, Louis Suarez, Alex Wang; Post-production and Drawing Contribution: Shenyuan Li, Kit Tang.

REFERENCES

Clifford, B., J. Lobdell, T. Swingle, and D. Zampini. 2019. "Walking Assembly: Craneless Tilt-Up Construction." In *ACADIA '19: Ubiquity and Autonomy, Projects Catalog of the 39th Annual Conference of the Association for Computer Aided Design in Architecture (ACADIA)*, edited by K. Bieg, D. Briscoe, and C. Odom. Austin, Texas: ACADIA. 50–55.

Cupkova, D., N. Azel, and C. Mondor. 2015. "EPIFLOW: Adaptive Analytical Design Framework for Resilient Urban Water Systems." In *Modelling Behaviour*, edited by M. Thomsen, M. Tamke, C. Gengnagel, B. Faircloth, and F. Scheurer. Cham: Springer. 419–431. https://doi.org/10.1007/978-3-319-24208-8_35.

Cupkova, D., and C. Clifford. 2018. "Moss Regimes: Embedded Biomass in Porous Ceramics." In *ACADIA '18: Recalibration: On Imprecision and Infidelity; Proceedings of the 38th Annual Conference of the Association*

of Computer Aided Design in Architecture, Projects Volume, edited by P. Anzalone, M. del Signore, and A.J. Wit. Mexico City: ACADIA. 96–101.

Cupkova, D., and M. Huber. 2021. "Rocking Cradle: Interactive urban furniture in pursuit of environmental attunement." In *Public Play Space Symposium*. Barcelona, Spain: Institute for Advanced Architecture of Catalonia. 132–139.

Craveiro, F., H. Almeida, H. Bártolo, P.J. Bártolo, and J.P. Duarte. 2017. "Topology and material optimization of architectural components." In *Challenges for Technology Innovation: An Agenda for the Future*. Lisbon, Portugal: CRC Press/Balkema. 71–76. https://doi.org/10.1201/9781315198101-14.

Dunn, Halpin. 2009. "Rugosity-based regional modeling of hard-bottom habitat." In *Marine Ecology Progress Series* 377: 1–11. https://doi.org/10.3354/meps07839.

ExOne. n.d. "Design Research Team Reinvents Eco-Friendly Architecture and Upcycled Materials." Accessed February 10, 2022. https://www.exone.com/en-US/Resources/News/CMU-Cradle-Case-Study.

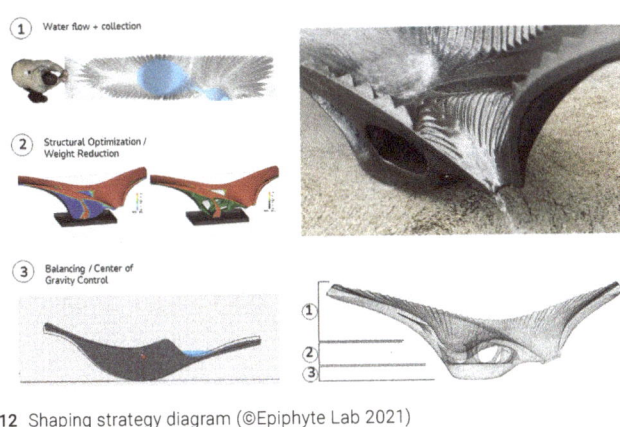

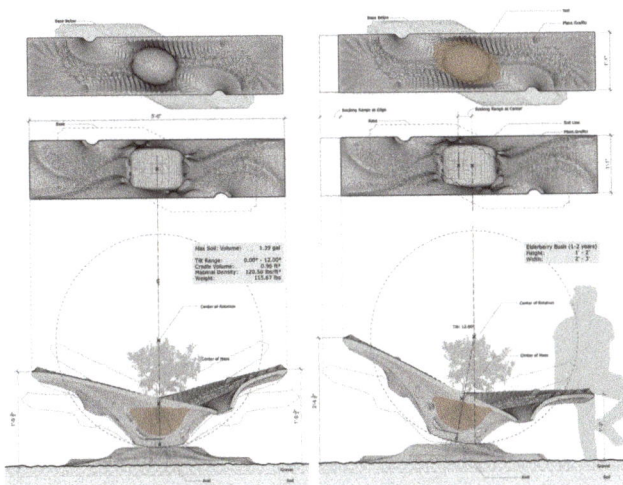

12 Shaping strategy diagram (©Epiphyte Lab 2021)

13 Drawing and computational analysis of a fabrication mesh model for *Planting Rocking Cradle* (©Epiphyte Lab 2022)

14 View at the *Water Collection Rocking Cradle* (©Massery Photography 2022)

Jeremy Faludi. J., C.M. Van Sice, Y. Shi, J. Bower, O.M.K. Brooks. 2019. "Novel materials can radically improve whole-system environmental impacts of additive manufacturing." In *Journal of Cleaner Production* 212: 1580–1590. https://doi.org/10.1016/j.jclepro.2018.12.017.

IMAGE CREDITS

Figure 1, 8 & 14: ©Massery Photography, Inc., 2022
Figure 11: ©Lake Lewis, 2021
Figures 3, 7: ©Arts Excursions Unlimited, 2020-21
Figures 2, 4-6, 8-10,12,13: ©Epiphyte Lab, 2020-22
All other drawings and images by the authors.

Dana Cupkova holds an Associate Professorship at Carnegie Mellon School of Architecture and directs Epiphyte Lab, Architectural Design + Research Collaborative. Engaging issues of environmental stewardship in design, Dana's work is situated at the intersection of built environment and ecology, focused on computational methods, materiality, embodied energy, and advanced manufacturing frameworks, with a particular interest in thermodynamics, and construction waste streams.

Matthew Huber holds the position of Special Faculty within the School of Architecture at Carnegie Mellon University, where he teaches issues of digital production, environmental ethics, building performance, and the influence of scientific practices and culture on architectural discourse. His previous experience in architectural practice involved developing projects simultaneously between large-scale planning with material and tectonic expression.

Edith Abeyta is a visual artist living in North Braddock, Pennsylvania. Focused on issues of equity and environmental justice, her works combine post-consumer goods, particularly clothing, and participatory gestures to form temporary installations and sculptures that explore collectivity, labor, and exchange. She frequently collaborates with other visual artists, poets, scholars, and the public. Edith leads the Arts Excursions Unlimited, a community-owned project dedicated to increasing the cultural connectivity of the citizens of the greater Hazelwood community.

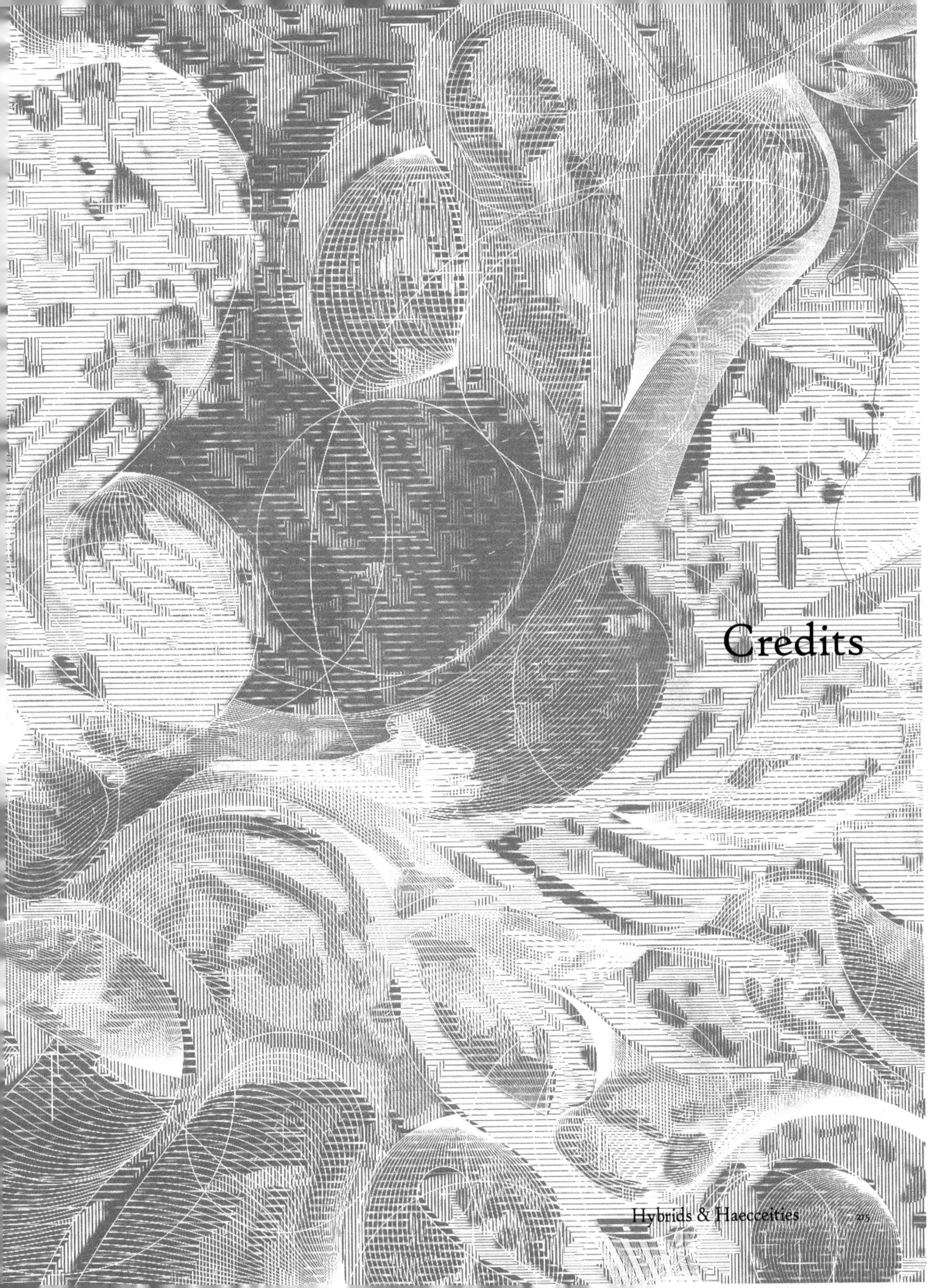

Credits

Conference Chairs

DR. MASOUD AKBARZADEH
Assistant Professor of Architecture, Weitzman School of Design, University of Pennsylvania

Masoud Akbarzadeh is a designer with academic background and experience in architectural design, computation, and structural engineering. He is Assistant Professor of Architecture focusing on Structures and Advanced Technologies and the Polyhedral Structures Laboratory (PSL) director. He holds a DSc from the Institute of Technology in Architecture, ETH Zurich where was a research assistant in the Block Research Group. In addition, he has two degrees from MIT: a Master of Science in Architecture Studies (Computation) and a MArch, the thesis for which earned him the renowned SOM award. He also has a degree in Earthquake Engineering and Dynamics of Structures from the Iran University of Science and Technology and a BS in Civil and Environmental Engineering. His main research topic is Three-Dimensional Graphical Statics, a novel geometric method of structural design in three dimensions. In 2020, he received the National Science Foundation CAREER Award to extend the methods of 3D/Polyhedral Graphic Statics for Education, Design, and Optimization of High-Performance Structures. He is also a Co-PI in a $4.6 million grant funded by National Science Foundation to investigate high-performance, self-morphing building blocks across scales toward a sustainable future. Has also received a $2.4 Million ARPA-E Grant to Research the Design of Carbon-Negative Buildings starting September 2022.

HINA JAMELLE
Associate Professor of Practice, Weitzman School of Design, University of Pennsylvania
Director, Contemporary Architecture Practice, NY SH

Hina Jamelle teaches final year Graduate Option Studios and directs the Graduate Program's Urban Housing Studios at the University of Pennsylvania Weitzman School of Design. She has held the Visiting Schaffer Practice Professorship at the University of Michigan. Jamelle is the co-director of the New York and Shanghai-based architectural firm Contemporary Architecture Practice and has co-edited issues of *Architectural Design AD* titled *IMPACT* (2020) as well as *Elegance* (2007). Hina Jamelle's book *UNDER PRESSURE* on urban housing was published in 2021. Founded in 1999, Contemporary Architecture Practice [CAP] is known for futuristic designs using digital techniques and the latest technologies for the design and manufacturing of architecture. Projects include commissions by The Museum of Modern Art [New York]; Reebok Shanghai, Lijia Smart Park, Chongqing, Wenjin Hotels, Beijing, NJCTTQ Pharmaceuticals, Nanjing, AMEC Technologies, Nanchang [China]; Samsung, Seoul [South Korea]; and IWI Orthodontics Clinic, Tokyo [Japan]. Contemporary Architecture Practice's projects have been exhibited extensively at the Museum of Modern Art, New York; the London, Beijing, and Shanghai Biennales; and the Tel Aviv Museum of Art, among others. They also have been featured in more than 250 major publications around the world. Co-Directors Rahim and Jamelle have won the Architectural Record Design Vanguard Award and were featured in Phaidon's *10x10x2* as one of the world's top 100 emerging architects. Their project, IWI Orthodontics in Tokyo, Japan was featured in Phaidon's *ROOM 100* as one of the most creative interior design projects of the century. In 2015 she was recognized as 50 Under 50 Innovators of the 21st Century by a distinguished jury.

DR. DORIT AVIV
Assistant Professor of Architecture, Weitzman School of Design, University of Pennsylvania

Dorit Aviv, PhD, AIA, is Assistant Professor of Architecture at the University of Pennsylvania's Weitzman School of Design, where she directs the Thermal Architecture Lab, an interdisciplinary laboratory focused on the intersection of thermodynamics, architectural design, and material science. Her work examines how architectural materials and forms can impact airflows, energy interactions, and human health. A recipient of a 2020 Holcim Award for Sustainable Design and Construction, Aviv's recent projects include a combined evaporative and radiative cooling prototype for desert climate, development of radiant cooling for hot-humid climates, a blockchain-enabled distributed environmental sensing network, and indoor environmental quality control and assessment technologies. She is currently working on a Department of Energy grant for designing carbon-negative buildings. Aviv holds a PhD in architectural technology (energy and computation track) from Princeton University, an MArch degree with a certificate in urban policy from Princeton University, and a BArch from The Cooper Union. She is a licensed architect and has practiced in design roles at Tod Williams Billie Tsien Architects, KPF, and Atelier Raimund Abraham. Her work was exhibited in the 2021 Venice Architecture Biennale, and she was the co-curator of the energy pavilion in the 2017 Seoul Biennale for Architecture and Urbanism. Her research papers have been published in leading scientific journals such as *Applied Energy, Indoor Air*, and *Energy and Buildings*.

ROBERT STUART-SMITH
Assistant Professor of Architecture, Weitzman School of Design, University of Pennsylvania

Robert Stuart-Smith is Program Director for the Masters of Science in Design: Robotics and Autonomous Systems degree (MSD-RAS), Assistant Professor of Architecture in the Weitzman School of Design, and an Affiliate Faculty member of Penn Engineering's GRASP Lab. He directs the Autonomous Manufacturing Lab in Penn's Department of Architecture (AML-PENN), and co-directs its sister lab in University College London's Department of Computer Science (AML-UCL). Stuart-Smith's research operates at the intersection of algorithmic design, robotic fabrication, and collective robotic construction—developing an integrated approach to design, manufacturing, and robot behavior through varying degrees of programmed autonomy. Stuart-Smith is Principal Investigator (PI) for the £1.2mil EPSRC project "Applied Off-site and On-site Collective Multi-Robot Autonomous Building Manufacturing" and Co-PI for a £2.9mil EPSRC research project into "Aerial Additive Building Manufacturing," involving collaborations with industry partners including Cemex, Skanska, Mace, Burohappold, Arup, MTC, Ultimaker, Kuka, and others. Stuart-Smith is a co-founder of the experimental research collaborative Kokkugia and architectural practice Robert Stuart-Smith Design. Prior to joining the faculty at Penn in 2017, Stuart-Smith was a Studio Course Master in the AA School's Design Research Laboratory (2009-17), and held visiting professorships at Washington University, RMIT, University of Innsbruck, among others. Stuart-Smith's work has been published in journals including *Nature, Science Robotics, AD Architectural Design*, and *Architecture D'Aujourd'hui*. He has lectured and presented in symposia at institutions including AA, Sci-Arc, CCA, MIT, RMIT, Angewandte, Strelka Institute, Tsinghua University, Texas A+M, and others.

Hybrids & Haecceities

Workshop Chair

ANDREW SAUNDERS
Director of Master of Architecture Program
Associate Professor of Architecture, University of Pennsylvania

Andrew Saunders is Associate Professor of Architecture at the University of Pennsylvania Weitzman School of Design and founding principal of Andrew Saunders Architecture + Design, an internationally published, award winning architecture, design, and research practice committed to the tailoring of innovative digital methodologies to provoke novel exchange and reassessment of the broader cultural context. The practice innovates at a number of scales ranging from product design, exhibition design, and residential and large-scale civic and cultural institutional design.

He received his Bachelor of Architecture from the University of Arkansas and a Masters in Architecture with Distinction from the Harvard Graduate School of Design. His current practice and research interests lie in computational geometry as it relates to aesthetics, emerging technology, fabrication, and performance. He has significant professional experience as project designer for Eisenman Architects, Leeser Architecture, and Preston Scott Cohen, Inc.

He has taught and guest-lectured at a variety of institutions, including Cooper Union and the Cranbrook Academy of Art, and, most recently, he was Assistant Professor of Architecture & Head of Graduate Studies at Rensselaer Polytechnic Institute in New York.

In 2004 he was awarded the SOM Research and Traveling Fellowship for Masters of Architecture to pursue his research on the relationship of equation-based geometries to early 20th century pioneers in reinforced concrete. His current practice and research interests lie in computational geometry as it relates to emerging technology, fabrication, and performance. He is currently working on a book using parametric modeling as an analysis tool of 17th century Italian Baroque architecture. Most recently Andrew won the ACADIA international fabrication competition for the production of the Luminescent Limacon. The design for this lighting fixture was inspired by Flemish baroque portraits of the Dutch ruff and builds on computational and material research from his seminar Equation-based Morphologies.

Exhibition & Media Chairs

FERDA KOLATAN
Associate Professor of Architecture, University of Pennsylvania

Ferda Kolatan is Associate Professor at the University of Pennsylvania Weitzman School of Design and the founding director of SU11 Architecture + Design. He received his Architectural Diploma from the RWTH Aachen and his M.S.AAD from Columbia University. SU11 is an internationally acclaimed practice based in Brooklyn, New York and is dedicated to the conceptual and material exploration at the intersection of contemporary culture, technology, and design. SU11's projects have been exhibited at renowned venues such as MoMA, FRAC Center, Walker Art Center, Vitra Design Museum, Art Basel, Artists Space NY, and SU11 has participated in the SIGGRAPH and ACADIA conferences, and the Venice, Beijing, and Istanbul Biennales. Ferda Kolatan has taught, lectured, and written extensively about architecture. In 2010 he co-authored the book *Meander: Variegating Architecture* with Jenny Sabin, and his new book *Misfits and Hybrids* is forthcoming in 2023. In 2016, his Penn Research-Studio on Cairo received the inaugural 2017 *ARCHITECT Magazine* Studio Prize. In 2011 Ferda was selected as a Young Society Leader by the American-Turkish Society in New York for his achievements as an educator and designer.

NATE HUME
Senior Lecturer, University of Pennsylvania

Nathan Hume is a licensed architect and principal of Hume Architecture. His design work and writings have been published in journals and periodicals including *Project, Log, Posit, Tarp, Paprika,* and *The New York Times*. Nathan has exhibited work in shows at The Druker Gallery, the A+D Museum, Yale Architecture Gallery, CAED Gallery, Land of Tomorrow, One Night Stand, and the New York Center for Architecture. He is a senior lecturer at The University of Pennsylvania and has previously taught at Yale University and Pratt Institute. He received a Bachelor of Architecture from The Ohio State University and a Master of Architecture from Yale University.

Departmental Chair

WINKA DUBBELDAM
Miller Professor and Chair of Architecture Department, University of Pennsylvania
Founder and Principal, Archi-Tectonics

Winka Dubbeldam, MArch MS-AAD, is a seasoned academic and design leader, serving as Chair and Miller Professor of Architecture at the University of Pennsylvania Stuart Weitzman School of Design, where she has gathered an international network of innovative research and design professionals. She also taught advanced architectural design studios at Columbia University and Harvard University, among other prestigious institutions. Dubbeldam was the External Examiner at the Architectural Association London (2006-2009) and is currently the External Examiner at the Bartlett UCL in London (2019-present). Professor Dubbeldam was named one of the *DesignIntelligence* 30 Most Admired Educators 2015. She has been a juror and chaired many international and national award juries, and has been keynote speaker at international conferences. Professor Dubbeldam is one of the creative directors for CityX for the Virtual Italian Pavilion at the 2021 Venice Architecture Biennale. Winka Dubbeldam is also the founder and partner of the WBE-certified firm Archi-Tectonics NYC LLC, widely known for their award-winning work, recognized as much for its design excellence as for its use of smart building systems, sustainable materials, and innovative structures. A recent book *Strange Objects, New Solids, and Massive Things*, was published by ACTAR Publishers, Spain, in the fall of 2021.

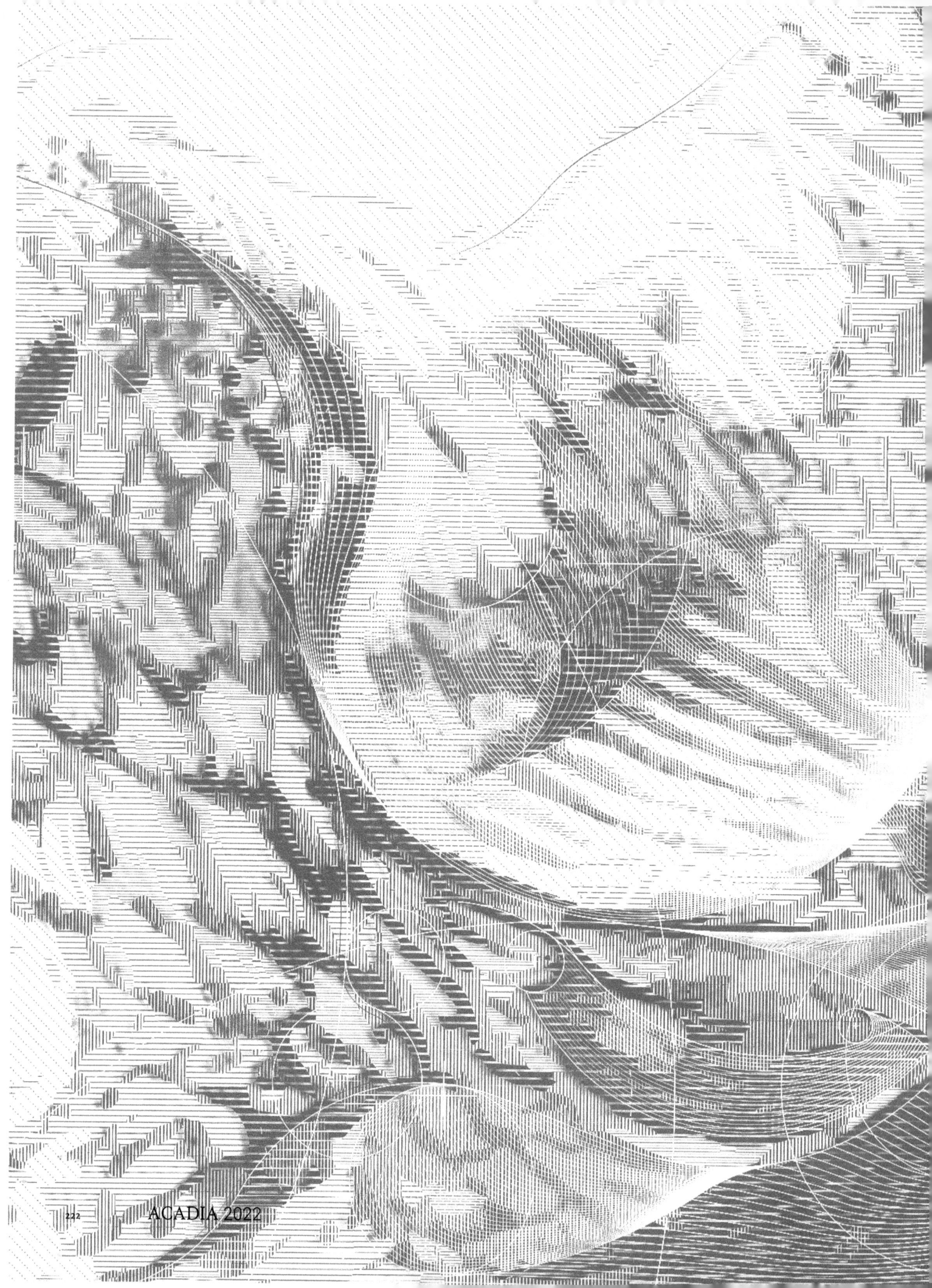

About ACADIA

The Association for Computer Aided Design in Architecture (ACADIA) is an international network of digital design researchers and professionals that facilitates critical investigations into the role of computation in architecture, planning, and building science, encouraging innovation in design creativity, sustainability, and education.

ACADIA was founded in 1981 by some of the pioneers of the field of design computation including Bill Mitchell, Chuck Eastman, and Chris Yessios. Since then, ACADIA has hosted over 40 conferences across North America and has grown into a strong network of academics and professionals in the design computation field.

Incorporated in the state of Delaware as a not-for-profit corporation, ACADIA is an all-volunteer organization governed by elected officers, an elected Board of Directors, and appointed ex-officio officers.

PRESIDENT
Jenny E. Sabin

VICE-PRESIDENT
Kathy Velikov

SECRETARY
Tsz Yan Ng

TREASURER
Phillip Anzalone

MEMBERSHIP OFFICER
Vernelle A. Noel

DEVELOPMENT OFFICER
Matias del Campo

COMMUNICATION OFFICER
Melissa Goldman

TECHNOLOGY OFFICER
Jose Luis Garcia del Castillo López

IJAC ACADIA OFFICER
Dana Cupkova

2021 ELECTION BOARD OF DIRECTORS
Term: January 1st, 2022 - December 31st, 2023

Shelby Doyle, *Iowa State University*
Behnaz Farahi, *California State University, Long Beach*
Maria Yablonina, *University of Toronto*
Leslie Lok, *Cornell University*
Kathrin Dorfler, *Technical University of Munich*
Sina Mostafavi, *TU Delft (alternate)*
Daniel Bolojan, *Florida Atlantic University (alternate)*
Leighton Beaman, *Cornell University (alternate)*

2020 ELECTION BOARD OF DIRECTORS
Term: January 1st, 2021 - December 31st, 2022

Matias del Campo, *University of Michigan*
Tsz Yan Ng, *University of Michigan*
Jose Luis Garcia del Castillo López, *Harvard University*
June A. Grant, *blink!LAB Architecture*
Stefana Parascho, *EPFL*
Biayna Bogosian, *Florida International University (alternate)*
Melissa Goldman, *University of Virginia (alternate)*
Vernelle A. A. Noel, *University of Florida (alternate)*

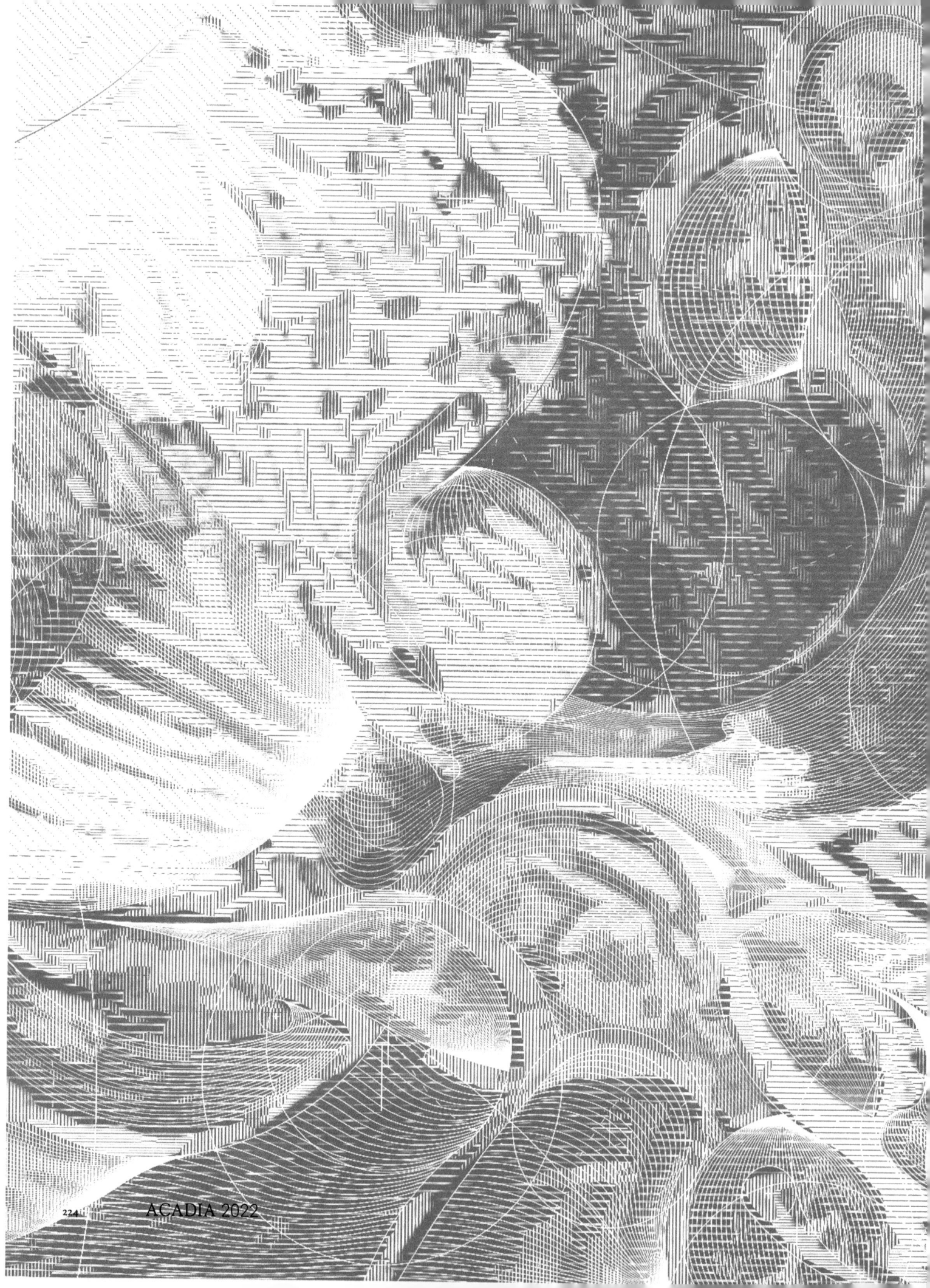

Conference Management

**WEITZMAN SCHOOL OF DESIGN
AT THE UNIVERSITY OF PENNSYLVANIA**

Fritz Steiner, *Dean and Professor*
Winka Dubbeldam, *Chair and Professor*

CONFERENCE CHAIRS
Dr. Masoud Akbarzadeh, *Assistant Professor of Architecture*
Dr. Dorit Aviv, *Assistant Professor of Architecture*
Hina Jamelle, *Associate Professor of Practice*
Robert Stuart-Smith, *Assistant Professor of Architecture*

WORKSHOP CHAIR
Andrew Saunders, *Associate Professor*

EXHIBITION & MEDIA CHAIRS
Ferda Kolatan, *Associate Professor*
Nate Hume, *Senior Lecturer*

WEBSITE & TECHNICAL
Christine Khouri Sader
Kevin Jed He
Leon Yi-Liang Ko

EXHIBITION
Lauren Hanson
Reem Abi Samra
Nicholas Houser
Joseph Depre

MEDIA
Jorge Couso

MERCHANDISE
Peik Shelton
Yasmin Goulding

GRAPHIC IDENTITY
Madison Green
Paul Germaine McCoy
Peik Shelton

GRAPHIC EDITORS
Anna Ji-Eun Lim
Mingyang Yuan
Hei Wai Valerie Tse

**WEITZMAN SCHOOL OF DESIGN ADMINISTRATION
AT THE UNIVERSITY OF PENNSYLVANIA**

Scott Loeffler, *Director of Administration*
Michael Grant, *Director of Communications*
Kait Ellis, *Executive Secretary to the Dean*
Hanna Finchler, *Associate Director of Communications*
John Caperton, *Communications Coordinator*
Christopher Cataldo, *Director of Finance*
Nadine Beauharnois, *Coordinator of Finance & Budget*

ACADIA STEERING COMMITTEE

Biayna Bogosian
Matias Del Campo
Shelby Doyle
Behnaz Farahi
Melissa Goldman
Sina Mostafavi
Cameron Nelson
Vernelle A. A. Noel
Jenny E. Sabin
Kathy Velikov

COPYEDITOR
Gabi Sarhos

Projects Peer Review Committee

Mostafa Akbari
University of Pennsylvania, Weitzman School of Design

Ali AlYousefi
University of Pennsylvania, Weitzman School of Design

Jeffrey Anderson
University of Pennsylvania, Weitzman School of Design

Phillip Anzalone
New York City College of Technology

Inés Ariza
ETH Zurich

Kristy Balliet
Southern California Institute of Architecture

Leighton Beaman
Cornell University College of Human Ecology

Mathias Bernhard
University of Pennsylvania, Weitzman School of Design

Kory Bieg
University of Texas, Austin School of Architecture

Yana Boeva
University of Stuttgart Institute for Social Sciences

Biayna Bogosian
University of Southern California

Ronan Bolaños
National Autonomous University of Mexico

Mohammad Bolhassani
City College of New York

William Braham
University of Pennsylvania, Weitzman School of Design

Johannes Braumann
Robots in Architecture

Danelle Briscoe
University of Texas, Austin School of Architecture

Nicholas Bruscia
University at Buffalo SUNY

Timo Carl
Frankfurt University of Applied Sciences

Jason Carlow
American University of Sharjah

Juan Jose Castellon
Rice University

Hua Chai
University of Pennsylvania, Weitzman School of Design

Mike Christenson
University of Minnesota

Greg Corso
Syracuse University

David Costanza
Cornell University

Dana Cupkova
Carnegie Mellon University

Mahesh Daas
Boston Architectural College

Pierluigi D'Acunto
Technical University of Munich

Patrick Danahy
Ball State University

Marcella Del Signore
New York Institute of Technology

Antonino Di Raimo
University of Portsmouth

Mark Donohue
California College of the Arts

Kathrin Dörfler
Technical University of Munich

Emre Erkal
Erkal Architects

Behnaz Farahi
California State University Long Beach

Richard Garber
University of Pennsylvania, Weitzman School of Design

Jose Luis García del Castillo y Lopez
Harvard University Graduate School of Design

Guy Gardner
University of Calgary

David Gerber
University of Southern California

Melissa Goldman
University of Virginia, School of Architecture

Rhys Goldstein
Autodesk Research

Marcelyn Gow
Southern California Institute of Architecture

Miaomiao Hou
University of Pennsylvania, Weitzman School of Design

Molly Hunker
Syracuse University

Nathaniel Jones
Arup

Negar Kalantar
California College of the Arts

Lydia Kallipoliti
Cooper Union

Neil Katz
Skidmore, Owings & Merrill LLP

Aysegul Akcay Kavakoglu
Istanbul Technical University

Ted Kesik
University of Toronto

Axel Kilian
Massachusetts Institute of Technology

Daniel Koehler
University of Texas, Austin School of Architecture

Ferda Kolatan
University of Pennsylvania, Weitzman School of Design

Axel Körner
University of Stuttgart

Sarah Aipra Kott-Tannenbaum
Quezada Architecture

Rodrigo Langarica
Universidad Anáhuac Querétaro

Christian Lange
University of Hong Kong

Carla Leitao
Rensselaer Polytechnic Institute

Marta Llor
Snarkitecture

Leslie Lok
Cornell University

Russell Loveridge
ETH Zurich NCCR Digital Fabrication

Ryan Vincent Manning
Quirkd33

Matan Mayer
IE University

Wes Mcgee
Taubman College of Architecture & Urban Planning

Frank Melendez
City College of New York

AnnaLisa Meyboom
University of British Columbia

Saurabh Mhatre
Harvard University, Graduate School of Design

Laia Mogas-Soldevila
University of Pennsylvania, Weitzman School of Design

Philippe Morel
University College London

Sina Mostafavi
University of Huddersfield

Kris Mun
ANFA

Taro Narahara
New Jersey Institute of Technology

Eduardo Sampaio Nardelli
Universidade Presbiteriana Mackenzie

Andrei Nejur
University of Montréal

Vernelle Noel
Georgia Tech University

Betul Orbey
Dogus University

Derya Gulec Ozer
Istanbul Technical University

Mine Özkar
Istanbul Technical University

Dimitris Papanikolaou
University of North Carolina

Stefana Parascho
CREATE Laboratory at Princeton University

Vera Parlac
New Jersey Institute of Technology

Vijay Pawar
University College London

Andrew Payne
Robert McNeel and Associates

Marshall Prado
University of Tennessee

Eleanor Pries
San Jose State University

Nick Puckett
OCAD University

Christopher Romano
University at Buffalo

Jose Sanchez
University of Michigan

Andrew Saunders
University of Pennsylvania, Weitzman School of Design

Anton Savov
ETH Zurich

Marc Aurel Schnabel
Victoria University of Wellington, New Zealand

Tobias Schwinn
University of Stuttgart Institute for Computational Design

Jane Scott
Newcastle University

Jason Scroggin
University of Kentucky College of Design

Brian Slocum
Universidad Iberoamericana

Kyle Steinfeld
University of California, Berkeley

Satoru Sugihara
Architectural Technology Laboratory Venture

Oana Taut
Institute for Advanced Architecture of Catalonia

Aron Temkin
Norwich University

Mette Ramsgaard Thomsen
Royal Academy of Fine Arts

Daniel Tish
Harvard University, Graduate School of Design

Kenneth Tracy
Singapore University of Technology and Design

Robert Trempe
Arkitektskolen Aarhus

Franca Trubiano
University of Pennsylvania, Weitzman School of Design

Richard Tursky
Ball State University

Kathy Velikov
Taubman College of Architecture & Urban Planning

Tom Verebes
New York Institute of Technology

Joshua Vermillion
University of Nevada

Zherui Wang
University of Pennsylvania, Weitzman School of Design

Nick Williams
Aurecon

Andrew John Wit
Tyler School of Art and Architecture

Jun Xiao
University of Pennsylvania, Weitzman School of Design

Shai Yeshayahu
Ryerson University

Lei Yu
Raytheon Technologies

Machi Zawidzki
Polish Academy of Sciences

Catty Dan Zhang
UNC Charlotte

Hao Zheng
University of Pennsylvania, Weitzman School of Design

Sasa Zivkovic
Cornell University

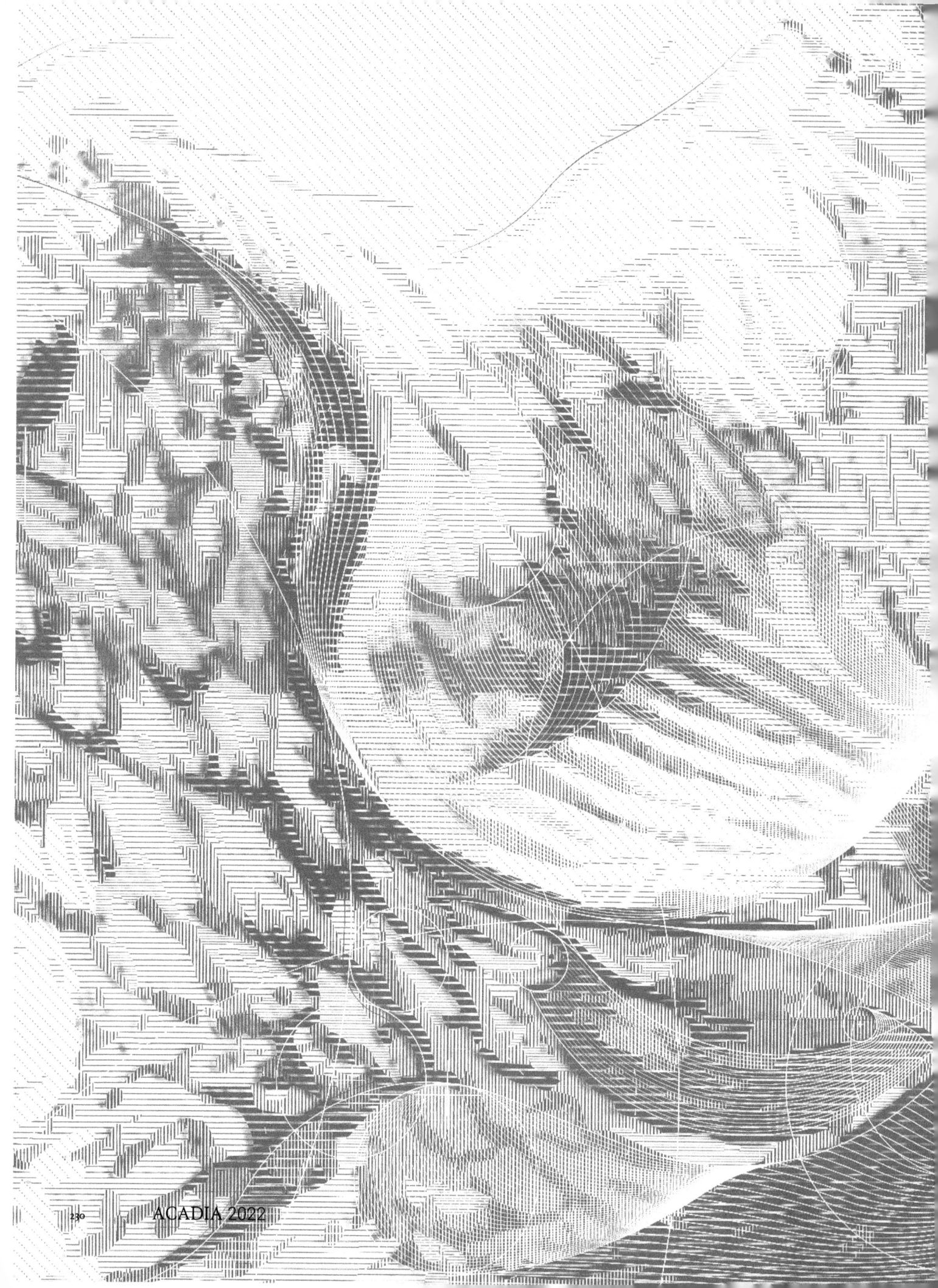

A Mosaic of Unlikely Affinities Credits

University of Pennsylvania, Weitzman School of Design

1 **Prototype for Ger, UNICEF, Mongolia**
Professor William Braham
PhD Candidate Evan Oskierko-Jeznacki, PhD Student Max
Hakkarainen

2, 3 **Global Innovation Center, Philadelphia**
Lecturer Scott Erdy
Erdy McHenry Architecture, LLC
Owner/Client: *Axalta/Liberty Pro*
Structural Engineer: *Bala Consulting Engineers, Inc.*
MEP: *Van Praet and Weisgerber*
Roof Consultant: *Tully International*
Civil: *LGA Engineering, Inc.*
Audio Visual: *Straub Audio Visual Systems*
Geo Technical: *Melick-Tully and Associates*

4 - 6 **Brooklyn Botanic Garden Visitor Center, Brooklyn**
Graham Professor of Practice Marion Weiss
WEISS/MANFREDI Architecture/Landscape/Urbanism

7 - 9 **Longwood Gardens, Brandywine Valley**
Graham Professor of Practice Marion Weiss
WEISS/MANFREDI Architecture/Landscape/Urbanism

10, 11 **497 GW Condominium, New York**
Miller Professor Winka Dubbeldam
Archi-Tectonics
Project Leader: Ana Sotrel
Project Team: Michael Hundsnurscher, Tanja Bitzer,
Deborah Kully, Nicola Bauman, Ty Tikari, Amy Farina
Architect of Record: *David Hotson, Architect*
Structural Engineers: *Buro Happold NY*
Mechanical Engineer: *Gabor Szakal Engineers*
Curtain Wall Consultant: Israel Burger, Bill Logan
Photography: *Floto & Warner Photography, Archi-Tectonics*

12 **Millennium Hall, Philadelphia**
Lecturer Scott Erdy
Erdy McHenry Architecture, LLC
Owner: *Drexel University*
Structural Engineer: *The Harman Group, ARUP*
MEP: *AKF Group*
Civil: *Pennoni*

13 - 15 **Hybrid Main Stadium, Hangzhou**
Miller Professor Winka Dubbeldam
Archi-Tectonics
Local Architect: *The Architectural Design & Research
Institute of Zhejiang University, ACRC*
Structural Engineer: *Thornton Tomassetti*
Transport & Egress Planning: *Mobility in Chain*
Landscape Design: *!melk NY*

16 **Lijia Smart Park Ford Mobility Center, Chongching**
Professor Ali Rahim and
Associate Professor of Practice Hina Jamelle
Contemporary Architecture Practice NY SH
Design Team: Caleb White, Leon Yi-Liang Ko, Yang Yang,
Chris Noh, Flori Kryethi

17 - 19 **Church of Saint Aloysius, Jackson**
Lecturer Scott Erdy
Erdy McHenry Architecture, LLC
Owner: *Archdiocese of Trenton*
Structural Engineer: *Bala Consulting Engineers, Inc.*
MEP: *Van Praet and Weisgerber*
Roof Consultant: *Tully International*
Civil: *LGA Engineering, Inc.*
Audio Visual: *Straub Audio Visual Systems*
Geo Technical: *Melick-Tully and Associates*

20, 21 **Asian Games Field Hockey Stadium, Hangzhou**
Miller Professor Winka Dubbeldam
Archi-Tectonics
Local Architect: *The Architectural Design & Research
Institute of Zhejiang University, ACRC*
Structural Engineer: *Thornton Tomassetti*
Transport & Egress Planning: *Mobility in Chain*
Landscape Design: *!melk NY*

22, 23 **Herzog & Loibner Jewelry Storefront, Vaduz, Liechtenstein**
Associate Professor Ferda Kolatan
SU11 Architecture + Design, NY
Partners in Charge: Ferda Kolatan, Erich Schoenenberger,
and Hart Marlow
Local Architect: *ID Connect Design Solutions*

Sponsorship

PLATINUM

Zaha Hadid Architects

SILVER

GRIMSHAW

BRONZE

EVENTSCAPE

SPONSOR

 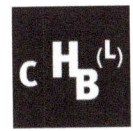 

MEDIA

The Architect's Newspaper

The Graphic Identity of Hybrids & Haecceities

Madison Green, Paul Germaine McCoy, Peik Shelton

The amalgamation of a hybrid and a haecceity is to recognize the existence of polarity and uniqueness while enabling relationships between such entities to simulate and blend qualities from one another which overtime appear to be "new" and "only" while exisiting in "multiplicities" and "varieties." The graphics proposal for the University of Pennsylvania Weitzman School of Design's hosting of the ACADIA 2022 conference titled "Hybrids & Haecceities" illustrates that things can exist as a thing in a state of transition or in between. In this way, lines that describe contemporary tools and discrete methods also outline familiar figures and forms. A graphics library extracts euclids from splines and hatches from colors to generate tweens of curvatures that—when sampled with AI learning—open up a relevant conversation of the computational, material, objective, aesthetic, genetic, biological, environmental, robotic, and operational to a new reading. A reading where the graphic aesthetic of the digital is informed by multiple layers of graphics and computer-aided design in architectural practice. This reading develops through constant curation, editing, tracing, adding, removing, fading, and overlaping of graphic elements. The compositions create dense overlay assemblies of rasterized and vector images. A library of cropped editions is created from this image to focus on different regions of information. Overall, Hybrids & Haecceities is a unique evolution of the familiar graphic reading of ACADIA.